Eleanor Morgan has written for the *Guardian*, the *Observer*, *The Times*, the *Independent*, *GQ*, *Harper's Bazaar*, *ELLE*, *Vogue*, *VICE*, *McSweeney's*, *The British Psychological Society's Research Digest* and others. Her first book, *Anxiety for Beginners: A Personal Investigation*, was published by Bluebird (PanMacmillan) in 2016. She subsequently completed a Master's in Psychology and is now training as a psychologist.

Also by this author

Anxiety for Beginners: A Personal Investigation

Hormonal

A Conversation About
Women's Bodies, Mental Health and
Why We Need To Be Heard

ELEANOR MORGAN

virago

VIRAGO

First published in Great Britain in 2019 by Virago Press

1 3 5 7 9 10 8 6 4 2

A CIP catalogue record for this book
is available from the British Library.

ISBN 978-0-349-01139-4

Typeset in Bembo by M Rules
Printed and bound in Great Britain by
Clays Ltd, Elcograf S.p.A.

Papers used by Virago are from well-managed forests
and other responsible sources.

Virago Press
An imprint of
Little, Brown Book Group
Carmelite House
50 Victoria Embankment
London EC4Y 0DZ

An Hachette UK Company
www.hachette.co.uk

www.virago.co.uk

For Kate

Contents

Part Three

Part Four

Part Five

Hormonal

Part One

Part One

That day

One day in our early lives as women, everything changes. We start bleeding. The beginning of our menstrual cycle, our fecundity, renders us different. In an instant we're no longer child. I was on a crazy golf course overlooking Cromer pier when I experienced period pain for the first time. It was the summer holidays and I'd started my period while away with my dad, brother and sister, just before my fourteenth birthday. I'd taken to wearing my hair all scraped back and my abundant forehead was absorbing the north Norfolk sun with vigour. As I stared at a pair of wooden clown lips I had to putt my ball between, my lower body rippled with new sensation. It was a pain that didn't fit inside my body. My pelvis hung suspended, like a bowling ball, threatening to burst from between my legs. That classic British coastal perfume, thick with sun-roasted kelp and old deep-fat-fryer oil, took the nausea that seemed to come with these sharp churns to another level. I had to sit down on the grass. I thought: this is *crazy*.

Girls at school talked about period pain. Some fainted in class or on the benches next to the netball courts because of it. One girl vomited all over her desk in a maths class and started crying before being led to the nurse's office, the teacher's hand gently holding the small of her back. These girls gained noble status. Their syncopes were crowns of maturity. To myself and others who hadn't yet 'started', they seemed a different species because their bodies knew things ours didn't. We non-starters used to ask each other how bad it could be, because we all *knew* what a stomach ache felt like. But for a stomach ache to have girls falling on their backs and puking into their pencil cases seemed wild.

Until I felt it.

This new pain burned through my thighs. My legs were like pipe cleaners as a teenager. That holiday they were so tanned they looked wood-stained. The blonde hairs on them caught in the light like fibreglass and, sitting on the grass, I wondered if I should start shaving above the knee (Mum had always said not to) like others did at school, rather than stopping at the cap. My thighs carried on burning. I explained to Dad that I didn't feel well. He nodded and told me to just 'sit quietly'. I was aware of the shift in his gaze towards me in a way that I absolutely did not have the language for. Only feeling.

Something set me apart from my younger sister. A vague sense of shame swam about me. A few days earlier we'd been in a four-man tent in Southwold and I'd sat by the zip feeling a peculiar nostalgia, or longing, that I couldn't place. I told myself I was homesick, even though home wasn't a particularly great place to be then. When we got to Cromer, I went to the toilet and found maturity in my knickers. An initial rush of excitement – I could go back to school and be part of

the in crowd! The swooning girls! – gave way to a funny sadness. I had to leave the house immediately after telling Dad. It
wasn't quite embarrassment that made me leg it – he did his
best to make it a non-thing; a mini-celebration, even – but
in my core I felt uncomfortable. That whole holiday was one
of heavy, wordless feelings.

I recall that afternoon on the golf course so clearly, I think,
because the overwriting of what my body once knew as pain
felt so significant at the time. The letters of the word *p, a, i, n*
are symbols. Abstract. But there was a physiological process
happening in my body and brain as it learned to accept this
new state and its corresponding language. I began to embody
the word differently. I had known pain before, of course.
Broken fingers, headaches, tonsillitis, bruises, scratches and
bite-marks from fighting with my siblings. This was different; it had texture. Emotion. I looked down at the grass and
thought, for a split second, that it would never stop. That the
pain was time itself.

Dad gave me the keys to go back to the house and take
some ibuprofen. I made cat-like sobbing noises the whole way.
Back at the house, I lay on the sofa waiting for the painkiller
to kick in. No one had mobile phones yet so I couldn't relay
my woes to any friends on WhatsApp. Anguished emojis
were years and years away. It was so quiet in that moment,
apart from the pitiful cawing of seagulls (why do they always
sound so desperate?) overhead. Strangely, though, it wasn't
lonely. The pain spoke to me. It told me my human fabric had
changed. As evidence of the encoding that happened that day,
now, on most occasions that I hear seagulls, I will have flash
visions of lying on that sofa holding my young, tortured belly.

I'm in my thirties now and, ever since my first period,
they've always been a slog. I've had my own swooning and

vomiting episodes as cramps ripped through my body; not ever at school, but often in shopping centres. I learned to deal with what always used to seem to me like an excessive amount of blood, which came in a variety of colours and consistencies. As a teenager, when I was 'on' (why did we say that? Upon what did we stand and then step off again?) I'd be forever looking backwards at myself in mirrors to check I hadn't leaked through my clothes: a steadfast paranoia that, although lessened, still lingers today.

I have spent quite some time now trying to gain autonomy over the way my hormones seem to make my mind and body behave each month. About five years ago a confidence-obliterating breakdown-of-sorts forced me to finally seek help for the anxiety that I had done my very best to conceal from everyone around me, *including* me, for well over a decade. At some point after that, in my ongoing quest for peace of mind, I started thinking about my menstrual health as part of my mental health. It was after a particularly bad period build-up one month, which felt like a week of tears-from-nowhere, overeating and spending too many evenings lying on my front in an existential torpor, that I went to my GP. I asked her if this could really 'just' be premenstrual syndrome (PMS).

Before the appointment I had somehow had the nous to start making a monthly diary (read: a series of symbols Sharpied on the Cliff Richard calendar my best friend bought me) and realised that this way of being and feeling crept up on me every couple of weeks, for a few days at a time. Notably, after my period, I'd have a week of feeling *almost* fantastic: productive, calm, resilient, porous to all the smell and colour of life. I say 'almost' because in the back of my mind I'd be worrying about feeling hijacked again. So I asked the doctor what I could do about feeling every single month like my soul was fermenting in a way I could not control.

Since then I have tried all sorts of interventions in my quest for emotional stability: more medications after several conversations with (almost exclusively male) gynaecologists that have made me feel either a) madder or b) ever so slightly less mad for a short amount of time; acupuncture; vitamin supplements; diet changes. Broadly speaking, this precise, reliable emotional state I sought was, is, elusive at best, impossible at most realistic. I will revisit my hormonal quest in much more detail later on; for now let's just say that few of the interventions brought me relief. In time, through addressing all the above with a new psychologist, and in the research I did for this book, I began to see things slightly differently and asked myself: what am I seeking relief *from*? Even if I could control certain symptoms of my premenstrual distress – low mood, for example – and a type of drug was the answer, on a much deeper level I wanted to know: what is the question?

One important aspect of this journey was discovering what I knew, didn't know or had forgotten, about what is going on inside my body: my baby-making machinery that, as yet, has made no babies, but for the love of god won't stop preparing for them each month. I wanted to explore in detail what happens at each stage during the menstrual cycle, because, from conversations I've had with many women, it seems that a greater awareness of the physical process can help us not only to manage our moods, but also to conceptualise what a mood actually *is*: a state of being that is, by its nature, temporary. Whether that awareness comes from using a period-tracking app, keeping a diary, doing online research or reading books, the hope is that we can develop a greater level of acceptance of our fluctuations in mood and emotion.

There are women out there who seem able to embrace their premenstrual selves without much shame because there are lots of different ways of being a woman. As I am typing this I am thinking of a brilliant line from an episode in season two of *The Good Fight*. Lawyer Diane Lockhart, actress Christine Baranski's multi-layered lead and also my hero, is accused by a young woman at the helm of a #MeToo-esque website called Assholes to Avoid of failing feminism for being part of the site's takedown. 'Women aren't just one thing,' she snaps, 'and you don't get to determine what they are.'

No woman gets to determine what another woman is, wants or desires.

In all the appointments I've had with GPs and gynaecologists I've been through a kind of re-education of what my body does and why. I have realised quite how little I remembered from my school education and felt quite ashamed by it. There's a flash vision I often have of sitting in a hall at secondary school with the rest of my year, watching a TV that was showing a cartoon woman inserting a tampon. Any anatomical accuracy was dispensed with in favour of a vague, hairless triangle that only flashed into view for a second. She appeared to take her trousers down, unwrap the tampon's packaging, position her body and push the thing up herself in one dainty movement. All with a big, shit-eating grin on her face. She was absolutely *delighted* about it all, our animated protagonist. Everyone giggled because, you know: *vaginas*. Obviously we learned more about sex hormones, puberty and reproduction in biology lessons (I even took biology as an A-level subject with a view of going to medical school) but until a few years ago when I had a sharp awakening regarding my fertility, I had forgotten a lot of it. Or had taken its function for granted.

In 2014, during an operation I was having to free up adhesions (bands of scar tissue) that were adhering my small bowel to my womb – part of the legacy of my burst appendix and all its gangrenous, bits-of-bowel-being-removed glory – the surgeon had received my consent to do a fertility test 'while they were there', like a bike mechanic might check your chain while repairing your brake cables. To do this they injected an iodine-containing dye into my womb to see if it came out the ends of both fallopian tubes. If it didn't, I might never be able to conceive without medical intervention. In recovery, as I was gingerly eating digestive biscuits and doing my best to pay attention through an opiate haze, the surgeon told me my fallopian tubes were damaged by scar tissue. I'd have to consider IVF if I wanted to become pregnant, or egg and/or embryo freezing. As the morphine flowered in my blood, I promptly vomited into one of those cardboard top hat things before the surgeon had finished talking.

After a year on the waiting list at my local NHS hospital, I embarked on the process of embryo freezing. With donor sperm bought from a bank in New York (I am a woman who sleeps with women: sperm is a little low on the ground around here), I ended up with five embryos ready to spend some time on ice. I think about them sitting in the freezer at my local teaching hospital often. During this gruelling process, I learned more from the doctors and nurses about what actually happens during the menstrual cycle, how my ovaries and hormones work, than I think I ever had before in my life. How bizarre that it took having fertility treatment to learn about myself, an adult woman who had been menstruating for nearly twenty years. Or was this awakening, of sorts, so unusual?

Through frank discussions I have had with other

women – family, friends, colleagues, research participants, sources for my journalistic work, strangers on social media – I have realised I am far from the only woman to have lived in a kind of disconnect from their body. We are everywhere. So at what point should we, like a London cabbie acquiring the 'Knowledge' of London's back alleys and one-way systems, have acquired the 'Knowledge' of our bodies? How could I have fallen so out of sync with what makes me *me*?

It feels like there is a movement happening regarding women's health right now. Menstrual-cycle tracking apps like Clue are becoming increasingly popular. Female entrepreneurs are creating new sanitary product businesses left, right and centre. More and more magazine articles are being written about previously 'icky' subjects like vaginal health, our menstrual cycles and the realities of menopause. We're talking more about the tyranny of ideals around birth control, conception, pregnancy and women's experiences of giving birth. We're talking about how the medical establishment doesn't always listen to our specific needs as women and how, when we're failed by healthcare, the effect on our sense of self can be catastrophic.

When I wrote an article for the website The Pool about how PMS was often poorly understood and routinely dismissed in modern healthcare, I received more messages from readers than I have about any article I've ever written. So many women shared the article with variations on 'thank god someone is saying it', and with details of how their own experiences had been diminished or stigmatised. In my day-to-day life I have noticed that in the last couple of years the women I know increasingly want to talk about aspects of their health they may have previously felt shy about. It is as though a thick, collective cork is being slowly winched out of us. Yet beneath the tentative conversations of liberation, the bottle is still full of mystery, ignorance and stigma regarding our bodies. It warrants our examination.

We must ask ourselves why, for example, cervical smear testing rates among young women have plummeted when case numbers of cervical cancer are increasing. When asked why they are reluctant to have a smear test, many young women speak about 'being judged'. The fear around having a smear test is not just about whether it will hurt when the speculum is put inside us, it is about how we feel our bodies will be perceived; whether our genitals look or smell 'right', for example. This gaze we cast on ourselves is an internalised male gaze – the product of not just the shiny, hair-free expectations of porn, but centuries of patriarchal oppression over our bodies in general. We are ashamed of what makes us women because deep down we believe that, in someone's eyes, it will never be quite right; that it will be too much.

The relationship we have with our body and what goes on inside it is deeply complex. Our reproductive systems are designed to grow babies and constantly prepare for them – even if we don't want them – and the hormonal fluctuations behind our reproductive processes mean that the workings of our inner world are often in collision with the outer world and all the expectations we perceive it to have of us. I am very interested in how society's continued perceptions of women affect our perceptions of ourselves. My personal experience of seeking treatment for PMS and for the sadness, heightened sensitivity and anxiety I can feel during my cycle, only to reach the point where I effectively ran out of treatment options, made me realise how entrenched my analysis of my own thoughts and behaviour was in more general ideas of how I 'should' or 'shouldn't' be as a person; a woman.

13

This book is underpinned by my own journey of becoming more knowledgeable about myself and accepting the emotional changes I feel over the course of my cycle, rather than seeing the changes as a pathology, forever scrutinising and labelling. It also became clear to me while researching the history of how society has understood, treated and talked about women's physical and mental health, that from the Hippocratic beginnings of medicine to the present day, our collective knowledge and attitudes have been shaped by the hubris of powerful men.

We know a lot more about what's going on inside us than we did in 460 BC, but we women have never quite lost the air of being mysterious creatures prone to physical and emotional excesses that, ultimately, need containing. When it comes to our bodies, in so many areas, our voice and our autonomy continue to be the weakest currencies. Someone always claims to know better. Medicine has an inherent bias against women. Women presenting at hospitals with pain are not only given fewer painkillers than men but are often offered sedatives instead (because we're easier when we're quieter). Treatment for women with coronary heart disease is delayed compared to men. All these facts are corroborated by robust data. When a woman says how she feels, what she wants or what she needs, someone – usually a man, or a system founded by a man – always knows better.

We can link women's biology as a source of oppression throughout history, and pejorative terms like 'hysterical', to the modern-day experience of women – both individual and collective – in so many ways. The massive societal reckoning that is happening in the wake of the #MeToo movement, for example, has led us to examine why so many women's horrific experiences of abuse have been disbelieved or minimised, and why so many women have felt powerless to speak out about something so painful. To me, the very concept of pain is key to why believing women is not a given, and to understanding our continued oppression. Because the reality is that, as women, our pain is not often taken at face value, and we know it.

Physical or emotional, the meaning of our pain has always been up for grabs, swallowed or dismissed by systems more powerful than us as individuals. Still, too, it is often the case that the louder we shout about our pain, the more we are told that we're causing trouble and the less likely we are to be taken seriously. Knowing *why* this is still the case in so many areas of society gives us something to work with and rally against.

This is a book about reclaiming meaning; about drawing a connecting thread between what history has told us about who and what we are, all the ways in which our so-called 'excesses' have been watched and contained, and the peace that can be found in trying to accept our inherent variability rather than forever striving to be or feel a certain way. A better understanding of what goes on inside us, of the connection between our bodies and minds, is an important part of that. So, too, is a greater awareness of all the external factors that may affect our bodily experiences, including the way modern medicine still doesn't take the variance of women's pain – or the voice describing that pain – seriously.

There is power in knowing ourselves better. We are not Post-Biology. We are protecting no one by pretending that we don't bleed; that our bodies, in preparing for and bearing children, don't make a mess. We are not reducing ourselves or undoing all the equality women have fought for by learning to embrace our messy selves. In fact, I'd say we're doing the opposite. So, let's start by going beneath the skin.

15

Intelligent flesh

It is impossible to be knowledgeable on every process happening inside us. There's just too much to comprehend. I start to feel a bit dizzy if I try to imagine it all: what my liver producing bile looks like, the wave-like contractions moving food along my gut, how the hot blood pumping around the 60,000-odd miles of arteries, vessels and capillaries would sound, if I could hear it. Unless we practise medicine or study the human body in a regular, rigorous way, we probably take for granted that every tiny chemical reaction in the body, every bacterial invasion that's fought away by our white blood cells, every process that keeps our internal conditions controlled – our water content, temperature and blood sugar levels, for example – is just *happening*, because of the innate intelligence of the body. Why wouldn't we?

Our skin and everything beneath it starts doing its job from the moment of our creation. It's enough, as an atheist, to make you understand faith in god. How does it all *know* what to do? Components of us go wrong, sometimes terribly, because every machine is fallible. But generally speaking, aided by what we feed and water them with, our factories of flesh keep us going, each bit doing its special job, until it all grinds to a halt one day. Have you ever felt a quiet awe watching a cut heal? At how, without conscious instruction, the skin starts regrowing almost instantly to prevent that vulnerable porthole to the inner world filling with bugs that could cause trouble? (I used to obsess over scabs when I was younger, looking at the lattice of clotted blood, like a tiny purple Shreddie, under a magnifying glass.) I never had any such reverence for my reproductive organs until I did the egg-freezing. The ones on the inside, anyway.

The female animal

Having a womb is a source of wonder for many of us, but it can also feel like a curse. The spectrum of potential pain and distress by virtue of having these organs is broad. Our reproductive systems are a kind of black magic, our wombs life-brewing cauldrons. But within the wonder of our nature is the reality that every woman, from puberty onwards, will experience some kind of physical and emotional turbulence around her reproductive system. As the writer Ariel Levy said in an interview to promote her magnificent book *The Rules Do Not Apply*, written after her visceral account of miscarrying appeared in the *New Yorker* ('Thanksgiving In Mongolia',[1] a piece of writing that affected me like nothing else I've read, before or since):

> Every woman is not going to decide to have children, every woman is certainly not going to lose a child, but at some point in her life almost every woman will have some kind of epic drama around menstruation, fertility, infertility, birth, menopause . . . something to do with this business of being a human female animal. It's part of life and it's not something that gets written about much. I felt like it was important to do that.

I feel like it's incredibly important to do that. This animalness of ours is one of society's last taboos. Discussions about the realities of periods, miscarriage, the viscera of birth, infertility, endometriosis or menopause are still, in twenty-first-century Western society, with all its discourse on equality and progression, often conducted in whispers behind the palm of a hand. Or not at all. This is in contrast to our conversations about mental health, which are becoming wider by the day. And yet these processes affect women's mental health in so many unignorable ways. If we are, gradually, getting to grips with the idea that mind is body and body is mind, and that, at its core, the term 'mental health' really equates to how we live and interact with others, what connections are we making about the patterns of mental distress in women?

In his 2013 book *The Stressed Sex: Uncovering the Truth About Men, Women and Mental Health*[2] Professor Daniel Freeman, a clinical psychologist at Oxford University, shared the results of twelve large-scale studies carried out across the world since the 1990s on gender patterns in mental health issues. Freeman found that women are up to 40 per cent more likely than men to develop mental health issues. His findings, based on analysis of epidemiological studies from the UK, US, Europe, Australia and New Zealand, also suggest that women are approximately 75 per cent more likely than men to report having recently suffered from depression and around 60 per cent more likely to report an anxiety disorder. Freeman's book garnered significant press attention and it was interesting to see how widely – and to what end – that 40 per cent statistic travelled. Simplistic headlines at the time suggested women were more 'at risk' of mental distress than men. Could it *really* be that we are inherently more vulnerable and, if so, why?

Freeman certainly wasn't suggesting anything as definitive. His study was large and, by looking at the general population, controlled for men being less likely to seek help for psychological problems than women, but the research is not a formal meta-analysis, which is when data from multiple studies are combined and analysed to get a more reliable estimate of data. There were no specific conclusions. There is not, as far as I can see from my searches of the literature, any robust evidence to explain a gender imbalance. So what we have is a body of evidence that tells us that women *might* suffer more, with no precise reason as to why. Perhaps, though, it's impossible to be precise.

Freeman said that a mix of factors was likely contributing to gender differences in mental health, relating to environment and societal factors as well as biological.

> Mental health issues are complex, they do arise from a range of factors, but we should highlight the environment, because we know discrepancies are greatest where the environment has the greatest role. Where we think it has an effect is particularly on women's self-esteem or self-worth: women tend to view themselves more negatively than men, and that is a vulnerability factor for many mental health problems.

This may be true. In all fields of brain study it is accepted that mental distress is not caused by one thing in isolation. The electrical, chemical brain is not extricable from the subjective mind. What we talk about is multi-causality: a fancy word for the concept that lots of different factors contribute to how we feel mentally. Therefore, when we talk about processes happening in a woman's body that place us at 'increased risk' of depression or anxiety, it seems not just blindingly logical but respectful to consider the ecosystem in which that body exists. Part of that ecosystem is a deep-seated stigma attached to the word 'hormones' but within it, too, is a disconnect between the lived realities of *being* a woman and what aspects of those realities are actually shared, explored and investigated.

19

As someone training in the field of psychology with a keen interest in women's health, I do believe that the biological bias of research in the field can be harmful. I believe, like Freeman, we should cast our attention to environmental factors. We nevertheless must not ignore the significant impact hormonal fluctuations can have on a woman's mental health across her lifespan. Biology *is* part of the picture.

The problem with discussing female biology is that it really has been stigmatised for so long. In the twenty-first-century Western world, it is sweet fantasy to say that all the myths, misinformation and ickiness surrounding female reproductive processes have been – finally – banished into the fusty past. It's a bit silly, really, because the word 'hormonal' applies as much to a testosterone-charged City banker as it does to a woman about to have her period. The metabolic processes of *all* organisms can only take place in very specific chemical environments and, in the human body, hormones (derived from the Greek ὁρμῶ, 'to set in motion, urge on') are special chemical messengers that, as part of the endocrine system, help to control most major bodily functions. Male, female, or anywhere on the spectrum of gender, hormones keep us alive. Yet this three-syllable adjective is still a tool for explaining away, with a shooing flick of the wrist, a woman's experience of her changing biochemistry. Men use it about women, women use it about other women and, of course, about themselves.

We blame so much on being 'hormonal': irrationality, 'bad' decision-making, angry outbursts, low self-esteem and other waverings in our mental wellbeing. *That wasn't the real me*, we might think, agonising over a knee-jerk decision we made at work, or taking offence at a throwaway comment someone made in the pub. *It was my hormones.* The same phrase could cover everything from wanting to kill your partner at the dinner table to having frantic sex with them over it. But what if there was a different way of looking at things? What if we could work towards dismantling the self-blame and, therefore, stop othering this fundamental part of who we are?

To do this, we must learn to become more aware and accepting of our animal selves. In light of how taboo aspects of our health still are, this is not a straightforward thing. I've always been one of those 'Come on, let's talk about our periods' women but, of course, respect that not every woman wants to – even though I'm interested in *why* she doesn't. Similarly, when I think about how disconnected I used to be about what happens inside me, how it can affect my state of mind, and how curious I then became about it, I respect that not every woman has the same curiosity. If you're able to ride the waves of your hormones without drowning, perhaps you wouldn't question what's happening. But there *is* something deep at play here. I have met so many women now who, like me, couldn't say with confidence what really happens, for instance, during ovulation. It seems strange that, whether or not we have children, our monthly cycle is a fundamental part of our lives for decades and decades – with no equivalent process for men – and that we can forget even the basics of what we're taught about ourselves at school. There has been a gradual sea-change, though. The increasing popularity of period-tracking apps like Clue, now offering more than just a digital period calendar and actually telling us what's going on at different points in our cycle, making predictions on how we might feel on any given day based on what data we have inputted in previous cycles, speaks of a growing curiosity; a desire to relearn ourselves. To listen to ourselves. Our mind–body balance is a synergy, but it can often feel like a confusing battleground. Even within, *especially* within, our own heads.

By first getting to grips with our biology a bit better, understanding the potential impact of our reproductive cycle, maybe we can then look at our lives and say, 'Okay, this week I might be more tired than usual and might want to take on fewer high-stress tasks, or do less intensive exercise,' or, 'For the next few days I might feel extra sensitive to the world and what people say to me, so I can maybe give myself less of a hard time when my emotional resilience seems to fall away.'

If what is happening in our bodies can have such an effect on our minds, should it be such a mystery to us?

Part Two

There is iron in her soul on those days.
She smells like a gun.

– Jeanette Winterson, *Written on the Body*

Part Two

Anatomy

To understand the menstrual cycle we need to first consider female anatomy.

We are born with two ovaries (where eggs are stored and released from), a womb (where a fertilised egg implants itself and develops into a baby), two fallopian tubes (thin, wiggly tubes that connect the ovaries to the womb), a cervix (the doorway to the womb from the vagina) and a vagina. From the front, the whole thing looks a bit like a ram's head: historically a pagan symbol for female sexuality. A brief peek into art history can give us some idea of our formative understanding of and views towards the female body: confused and, often, disdaining.

In 2016 a group of Brazilian academics claimed that Michelangelo had hidden references to the female reproductive system all over the ceiling of the Sistine Chapel in his 'Creation of Adam' fresco. Their paper in the journal *Clinical Anatomy* observes that the recurring image of a ram's skull and horns closely matches the anatomy of a uterus.[3] They argue that, with his discreet anatomical allusions, Michelangelo was attacking Catholic misogyny. (His male contemporaries spent a lot of time arguing over whether women had souls or not.) However, as the art critic Jonathan Jones countered in the *Guardian* at the time,[4] 'Michelangelo had no interest in women or their bodies. He is the very least likely candidate to be a feminist artist . . . It is a cliché but undeniably true that when Michelangelo portrays nude women he very obviously portrays a male body, then clumsily sticks on a couple of marble breasts.'

Michelangelo may not have been the feminist hero the Brazilian researchers say he was, but he is known to have completed anatomical dissections of human bodies in order to paint and sculpt them with more accuracy. He may have dissected many women along the way, but left no sketches behind of his findings. Leonardo da Vinci, on the other hand, one of history's greatest artists and a contemporary of Michelangelo, did.

Leonardo's stunning anatomical drawings were centuries ahead of their time. His greatest triumph was understanding and recording to paper the mechanics of the human heart and blood circulation more than a century before any formal scientist got close to doing so. He melted wax and injected it into an ox's heart to make a cast, made a glass model of it and filled it full of water so he could watch how the vortexes worked. His conclusion that a swelling at the root of the aorta made the aortic valve shut after every heartbeat was one cardiologists did not confirm until the 1980s.[5] In short, the man was a mastermind. Here's where it gets interesting, though: Leonardo is also thought to be the first person in history to accurately portray a human foetus in its correct position inside a woman's womb, a strangely heart-wrenching tangle of little limbs, as well as the first to properly draw the vascular system of the cervix, vagina and womb.[6] On the down side, in one of his drawings of a foetus within the womb he incorrectly included the comb-like structures of the uterine walls he had seen when dissecting a cow. Among Leonardo's prodigious scientific discoveries, then, was the apparent interchangeability of bovine and human flesh: one of several (forgivable – this was over five hundred years ago) misunderstandings of the female body. Today, we might understand it a whole lot better, but the tang of historical stigma still hangs in the air. We'll get into this later.

As women, our own knowledge of what is inside us and what each part does may begin with conversations we have with our parents, other family members or friends. Any formal education usually begins at school, with classes on puberty. We tend to learn about puberty as we're either approaching it or in its sweaty throes. My school education happened in the UK in the 1980s and 90s. I learned about periods, sex, having babies and the whole how-to-prevent-having-them-if you-don't-want-them-yet thing in primary school (around the age of ten) and then at various points throughout my secondary education (between eleven and sixteen). The idea behind putting sex education on the curriculum, you hope and imagine, is that we develop a baseline knowledge of what we're made up of, what the bits we're made up of do and how they serve to continue our species. (All of the related safety, consent and emotional aspects are usually taught separately – to what degree and with what level of enthusiasm is highly variable.) Like the details of the Abyssinian Crisis from history lessons, though, it all got a bit foggy as I got older. Allow me to jog your memory.

Puberty

All the components needed to go through puberty are present from when we're born but our bodies keep them switched off for quite a long time. The average age for girls to begin puberty is eleven;[7] however some girls start as early as nine and some as late as sixteen. Usually, the first sign of puberty in girls is that the breasts start to develop. I remember my nipples sort of popping out one day, like two little foam sweeties. They were unbelievably tender, as is very common. Sometimes, one breast 'bud' starts to develop months before the other – again, very common. Pubic and underarm hair starts to grow. Over the next couple of years the breasts continue to grow and get fuller in shape. We begin sweating more, necessitating the use of deodorant, unless we actively want to walk around smelling like a pile of cut onions. The hormonal changes we go through stimulate the sebaceous glands to make more sebum and the glands become over-active, often resulting in spots. (My forehead bore similarity to the images captured by NASA's Mars Curiosity rover on the Red Planet.) We grow taller and gain weight as our body shape changes. Our hips get rounder and more body fat develops around our thighs, upper back and arms – all stimulated by oestrogen. At some point, usually a couple of years after puberty has begun, our periods start. Along with all the fun they entail. In the scientific world, our first period is referred to as 'menarche'. In Western society the average age for menarche is thirteen.[8]

Did you know that only humans, closely related primates, some species of bat and elephant shrews visibly menstruate?[9] I wonder if it would be possible to manufacture sanitary protection small enough to fit an elephant shrew. Two-millimetre-by-two-millimetre sanitary towels, maybe. In any case, for humans, the periods beginning means that the hormones previously held in check throughout our early childhood have been switched on. A small but very impor-tant part of the brain called the hypothalamus, responsible for linking our endocrine and nervous systems, starts to release regular, large bursts of gonadotrophin-releasing hormone (GnRH). This in turn stimulates the pituitary gland, another small but important part of the brain (about the size of a pea, living behind the bridge of our nose) that's often called 'the master gland' because it controls several other hormone glands in the body, including the adrenal and thyroid glands, as well as the ovaries and testicles. The pituitary gland begins producing luteinising hormone (LH) and follicle-stimulating hormone (FSH), which in turn cause a girl's ovaries to start producing *more* hormones, the ones we hear the most about: oestrogen and progesterone. These are known as sex steroids. FSH, LH, oestrogen and progesterone all play a part in regulating our menstrual cycle.

At the beginning of our reproductive years it's very common not just for our cycles to be irregular, but for the menstrual experience itself to be quite different month to month. The first period can be quite short with minimal bleeding, while the second may be longer with a heavier, more painful flow. Sometimes it may even seem to skip a month and return unexpectedly. The cycle becomes more regular after a couple of years and, within about six years of starting our periods, we usually settle into a predictable rhythm.

These initial years of irregularity can be attributed to the fact that our hormones are trying to balance themselves. They need to go up and down in level over the course of each cycle to fulfil their functions, but, during the first years of bleeding, the fluctuations are not quite regular enough to trigger ovulation in every cycle. If ovulation has not occurred, it is termed an anovulatory cycle. In a normal cycle, the release of the egg from the ovary stimulates the production of progesterone. In an anovulatory cycle, in which an egg has not been released from the ovary, too little progesterone is produced. This can have several results: our period not coming when we expect it, heavier bleeding and a higher chance of painful cramps. Anovulatory cycles are also very common in women who are approaching the menopause, when our hormone levels change dramatically again.

The ovaries also start to produce a certain amount of the male sex steroid, testosterone, during puberty. This hormone is important for women in its role of promoting bone and muscle strength, energy levels, mental wellbeing and libido. Testosterone is also responsible for the sexual sensitivity of our nipples and clitoris.

I'd like to take this opportunity to thank it personally.

Eggs and fertility

The ovaries may be small, looking a bit like a pair of new potatoes suspended in the lower abdomen, but they are powerhouses. They don't just produce hormones, they contain *hundreds of thousands* of eggs. We are born with them all; a staggering fact when you consider how small the ovaries are at birth. Your mother was born with all her eggs, which means your grandmother, when she was pregnant, was carrying a part of what became you inside her. It's incredible. Obviously, we are not going to have hundreds of thousands of babies, but nature gives us a big back-up system, because no new eggs develop after we enter this world. In fact, our 'ovarian reserve' – the number of viable eggs a woman has in her ovaries – declines from conception to menopause. A study led by by Dr Hamish Wallace of the University of St Andrews and Edinburgh University in 2010 was the first to actually collate this decline. Wallace's research showed that, on average, women are born with 300,000 potential egg cells, but this pool decreases at a much faster rate than previously thought.[10]

Wallace's model showed that, for 95 per cent of women, only 12 per cent of the maximum ovarian reserve is present by the age of thirty. Only 3 per cent remains at forty. At the time it was published, various sources claimed this research was 'the latest to warn women that they must not leave it too late to conceive' (*The Telegraph*),[11] but what the study (with its relatively small sample size of 325) also showed was how enormous the difference can be between individual women's ovarian reserve. Some women had as little as 35,000, others more than two million. Although it is extremely prudent to consider the fertility drop we have in our mid-thirties, it is also true that women *do* have children throughout their thirties and into their forties. Even fifties. If you are referred for fertility treatment like IVF because you are struggling to conceive, a specialist will investigate your ovarian reserve to predict how well your ovaries might respond to the treatment.

If we are contemplating having a baby, wondering whether we *can* have one or are actively trying, the conveyor belt of news stories regarding fertility just adds to the uncertainty so many of us carry. The discourse is cacophonous. Whether it's about how women choose to try to gain a sense of autonomy over the ticking of their biological clocks, or what it 'means' to become a parent in your forties, facts and opinions swarm like bees. Sometimes it can be helpful to strip things back to basics and work from there.

By definition, 'fertility' refers to a person's ability to produce offspring. For women, this commonly concerns ovulation: the monthly release of an egg. For men, it's the quality of semen: the fluid containing sperm that is ejaculated during sex. Women have what's known as a 'fertile window' each month, a day or two either side of ovulation, although, as we're taught in biology lessons at school, we can get pregnant at any time during our cycle if we're having sex with a man without contraception. Millions of us are now using cycle-tracking apps and know roughly when we're ovulating. I use Clue, but am in a same-sex relationship. I don't have the kind of sex that requires being aware of optimal pregnancy conditions; I track my cycle to be in tune with my hormonal fluctuations and their significant effect on my mood.

Lots of women having sex with men use fertility-tracking-based contraception apps like Daysy and Natural Cycles and have, despite doing what they say on the tin, fallen pregnant. Such apps are, as the journalist Dawn Foster wrote in the *Guardian*, 'simply the Vatican-favoured rhythm method repackaged in shiny, Silicon Valley jargon and a slick interface'.[12] They have attracted considerable scrutiny. A remarkable paper by a scientist called Chelsea Polis was published in 2018, showing how fatally flawed the analysis supporting Daysy's effectiveness was. These apps offer a kind of medical intervention. If they are making health claims, they must bring good evidence. As Polis argues in her paper, 'marketing materials on contraceptive effectiveness should be subjected to appropriate oversight'.[13]

In order for two humans to reproduce, a sperm cell must enter an egg cell so their genetic information can merge. The fertilised egg that results is called a zygote. Cell division then begins and the dividing zygote gets pushed along the fallopian tube towards the womb. Around four days after fertilisation, the zygote has about 100 cells and is called a blastocyst. When the blastocyst reaches the womb lining, it bobs around for a

couple of days before nestling itself into the uterine wall –
usually six or so days after fertilisation. This is the beginning
of pregnancy.

The sperm-and-egg reproductive basics are similarly
binary almost everywhere else in the animal kingdom. Some
species are known to be completely asexual, however, and
have no need for a male to reproduce – some species of whip-
tail lizards, for example. There are also some animals that are
designed to make babies with a male but don't always choose
to. The phenomenon of parthenogenesis, or 'virgin births', is
as spectacular as it is unnerving. In 2015, 'virgin-born' saw-
fish were seen for the first time in the world. The same year,
a study published in the journal *Animal Behaviour*[14] suggested
that female spiny leaf stick insects might prefer to embark on
parenthood alone because sex with males can – make your
own comparisons – be damaging to them. All partheno-
genetic offspring are female, too, so if the females carry on
going it alone, the males could be effectively wiped out. The
female insects have been known to fight off amorous males,
first squirting them with an anti-aphrodisiac chemical, then,
if they won't back off, aggressively kicking their legs. I curl
my fists in frustration at not being able to interview one of
these sticky iconoclasts.

We female humans cannot make babies by ourselves, alas,
much as some of us might like to. However, how an embryo
ends up in a woman's womb, and indeed where the constit-
uent parts of that embryo came from or when it was 'made',
is, thanks to advanced technology, a variable thing. We have
options. Donor sperm. IUI. IVF. Speaking as a woman who
imagines children in her future and also falls in love with
other women, those options are important to me. But the
beginning of a new human life can only, for want of a better

phrase, come about one way: as unavoidable a fact as our reproductive system's use-by date, constantly shone in our eyes like an optician's torch.

Nature doesn't always follow its own rules, however. Women who have been unequivocally told by doctors they cannot have kids end up getting pregnant. Some of my friends are in this category. Many women worry about whether they'll be able to have children in the future, particularly if they are living with conditions known to potentially affect fertility, like PCOS or endometriosis. Often, until we start trying for babies, we have no idea how easy or hard it will be. If we do find it hard to conceive, we might wish we'd started looking into things earlier. These concerns lead many women to have what is known as a 'fertility MOT': a blood test that measures a hormone known as Anti-Mullerian Hormone (AMH), which gives an idea about a woman's ovarian reserve.

For women who feel out of control regarding their fertility, this MOT offers a delicious sense of reassurance. However, the test has been reported as 'a waste of time and money' by the NHS, costs hundreds of pounds and, many doctors argue, is a way for pricey fertility clinics to exploit women's fears. The criticism seems justified, especially given that the blood test cannot assess the *quality* of your eggs – the most important part of the picture and something that, again, we don't really know about until we start trying for a baby.

An American study published in 2017[15] found fertility MOTs did not accurately predict a woman's chance of conceiving. Instead, the results showed that it doesn't matter how many eggs a woman has in reserve; the crucial factor is whether she's still releasing eggs (ovulating) regularly. So, just because your MOT results come back 'normal', there's no guarantee you'll be able to conceive a baby. Equally, an 'abnormal' MOT doesn't mean you won't be able to get pregnant. Given the evidence, it is not merely cynicism to suggest that Harley Street doctors are profiting from our anxiety, selling us a compelling-sounding 'fertility check' that doesn't hold up.

We are sold fertility reassurance by rich businesses *all* the time. Egg-freezing is another contentious subject. Barely a week goes by without another headline relating to the process. I should know. I wrote about my own experience for the *Guardian* in 2015 and have been asked pretty much every month since by a different TV producer to speak on a panel when a new perspective or study is published. The last one was with Katie Hopkins and they seemed surprised that I said no. But there is something big happening out there.

The first official report on egg-freezing in the UK from the Human Fertilisation and Embryology Authority (HFEA) shows an astonishing increase – 460 per cent – in the number of women freezing their eggs since 2010.[16] The report also shows that, despite the surge, egg-freezing cycles still only account for a minute 1.5 per cent of UK fertility treatments. In 2016, only 19 per cent of egg-freezing cycles were funded by the NHS, for medical reasons such as preserving the fertility of women who were undergoing chemotherapy or, like me, are infertile due to organic cause. The other 81 per cent were carried out in private clinics and were likely down to what are referred to as 'social' reasons, i.e. not having a stable partner or having financial or professional concerns. To deem the reasoning of a woman who isn't ready to be a parent just yet but who is aware of her declining fertility as 'social' is pejorative to say the least. So too is the current law that dictates how long a woman can freeze her eggs for if she has done so for non-medical reasons.

Fertility experts are, thankfully, now urging the government to dump the legislation that requires women who have frozen their eggs to use them within ten years. When the ten years are up, fertility clinics are obliged to destroy the eggs, irrespective of what the woman they belong to wants – unless she has been through the egg-freezing process because her fertility is compromised. On the most basic level, how can that be right?

As Baroness Ruth Deech, the British bioethicist and politician who was chair of the HFEA from 1994 to 2002, and who is one of the most vocal campaigners for a change in this law, says, the limit is 'arbitrary' and not in keeping with the current technology that is capable of keeping eggs safely frozen indefinitely. In an interview with BBC Two's Victoria Derbyshire, Deech said a change in legislation would 'be very easy to do, wouldn't cost anything and would give hope to women. There is nothing medically wrong with it, and I simply cannot see why the government won't give this attention.' The government is, in fact, disregarding how this current law breaches human rights law. But why consider the paltry matter of human rights when you're afraid of upsetting your hoary backbenchers? Recalling a meeting with health minister Jackie Doyle-Price, Deech said, 'I think she was frightened the anti-abortionists would pile in.' This law must change not just to remain in line with technological advances, but to stop treating women's hopes and desires as something so ephemeral. We can't stop our biological clocks ticking, but god knows we deserve the option to take the reins if we choose to.

We cannot ignore how commercialised egg-freezing has become, either. Eighty-one per cent of cycles in 2016 were carried out in private clinics, and set the women back, on average, £3,350 a go. These women are not stupid. As those researching the phenomenon have pointed out, the process is often entered upon with clear-eyed realism. It is true that improvements in egg-freezing technologies have offered women a far better chance of egg survival now. And it is quite wrong that women who have not found a partner should be penalised with childlessness, or pressured into finding one because their fertility is declining.

Egg-freezing extends the window of opportunity, but it can also be very taxing, mentally and physically, requiring hormonal stimulation of the ovaries by daily injections and surgical retrieval of the eggs under anaesthetic. It is IVF, just without the embryo implantation at the end. Success rates are not all that high and, while women embarking on the process should absolutely be supported in their decision, they should also be made aware of how realistic their chances of future pregnancy are. Regulating bodies must start to engage more fully with questions around the growing commercialisation of this process.

Our choices and attitudes towards baby-making are in addition loaded with cultural and societal meanings. Hierarchies. Embedded in the layers is abundant stigma and shame. Recent research suggests that many women may have gone through invasive processes because of the lack of historical focus on male fertility. On the face of it, this could be telling us something about whose feelings we've protected more.

All this eggy talk makes me think of salmon bellies, bulbous and heaving with shiny amber roe. Tiny, slippery pearls being spooned onto blinis. I told my own fertility specialist this once and he looked at me with the same empty stare I saw when he was about to insert the vaginal ultrasound probe and I asked him whether or not he thought he should buy me dinner first. I have a serious problem with not doing a 'bit' in medically intimate situations. We have evolved from fish, though, so frankly I think this man was impolite to not acknowledge my observation.

There are aspects of human anatomy that can only really be decoded by our marine ancestry. Human embryos, in their early stages, look eerily similar to the embryos of every other mammal, amphibian or bird, and we have all evolved from fish. All mammal eyes start out on the side of the head then move towards the middle. Humans start with gill-like formations on the neck that become the palate and jaw. However, unlike creatures such as sharks who have their reproductive organs – known medically as the 'gonads' – all the way up in their chests, ours start there but need to drop down. In men they descend and become the testes. In women, they descend and become the ovaries. It is astonishing, really, the different stages of growth that tiny embryo goes through in the womb before it becomes recognisably human.

Blood

In this monthly process we go through to prepare our body for a potential embryo, the ovary selects one egg to become 'ripe' (mature) and will release it to travel down the fallopian tube and into the womb. This is when sperm cells are supposed to swim inside in their millions, fighting with one another for the privilege of fertilising the egg. The rise in oestrogen in the first part of the cycle has already caused the womb lining (endometrium) to thicken and fill with blood in preparation for a pregnancy. After ovulation, progesterone rises and prepares the endometrium to become a comfy and hospitable place for a fertilised egg to grow into a foetus. If no sperm are invited in and the egg is not fertilised, the levels of oestrogen and progesterone produced by the ovary start to drop and, without their supportive function, the blood-rich endometrium is shed through the cervix and out through the vagina.

This is our period: a complex fluid made up of blood, vaginal secretions and the cells of that thickened endometrium. Rarely is the period a steady flow of pure red liquid, like the blood that seeps from our skin if we cut ourselves shaving. It is very normal to see clots when bleeding is heaviest, usually in the first couple of days of the period. They sometimes look like offal-y clumps of flesh or have a jelly-like consistency, and vary in number and size. (The first time I saw one on a piece of toilet paper I almost fainted.) Coagulation – the clotting process that turns fluid blood into a solid or semi-solid state – is the response our body uses to halt bleeding from an injury: very necessary for our survival. It would, however, cause problems if our menstrual bleeding were halted partway. Most research in this area concurs that our bodies release anticoagulants to prevent the menstrual blood from clotting as it is being expelled, but that, when the flow is heavy, these anticoagulants don't always have enough time to start working. Accordingly, we pass clots.

No one really tells you about the clots, do they? The bits of stuff that float to the surface of the bath if you lie in it without a tampon in, like frayed-looking shreds of a vital organ. At least, no one told me about the true nature of period blood in lessons at school. Not that I can remember, anyway. My mum never did either. I was obsessed with the little blue fold-up leaflet in her boxes of Lil-Lets, but only ever recall reading that women lose 'around an eggcupful' of blood each month. To this day I think: *Pardon?* In the first few years of bleeding

I'd so often be shocked, looking at it in the toilet below me, wondering: *How* can it be normal that this, a crimson Cy Twombly on the porcelain, has come out of my body and I'm not about to die?

That incredulity mostly goes away. Cleaning up blood becomes quotidian. However, I started using a menstrual cup recently and was quite taken aback at what the contents looked like when I first removed it (not the easiest of tasks). Like a tiny trifle, it had *layers*. On top of the red blood, itself above a sediment of dark-looking tissue, was a layer of clear, yellowish liquid. I must have stared at it, held slightly aloft over the sink, for five minutes. Tampons and towels absorb so much of what leaves our bodies each month. I'd never imagined it would look quite like this. After some extensive googling I discerned that the top layer was a mixture of the normal cervical mucus and the clear plasma separating from the red blood cells. (Isn't plasma one of the nicest words to say out loud?) It all separates out after a while when it's left to sit upright in your body. Moving around a lot, cartwheels and the like, will keep it all mixed up; our vaginas a veritable cocktail-shaker.

The first few months of using this tiny silicone cup gave me both a sense of pure, childlike wonder at a bodily fluid and a greater admiration for my body itself: working, preparing, always trying again. I'm still occasionally dazzled by how urgent it looks. Sacrificial, almost. When I'm emptying the cup I often think of that famous passage in Stephen King's *Carrie*. His first novel, written in 1974, it contains probably one of the best-known depictions of menstruation in pop culture. When its protagonist gets her first period in the communal showers at school, she hasn't been taught about periods, puberty or anything of the body, really, due to her mother's fanatical faith and the reign of her abusive father. So, when she starts bleeding, she obviously thinks she's dying.

'For God's sake, Carrie, you got your period! ... Clean yourself up!'

Carrie backed into the side of one of the four large shower compartments and slowly collapsed into a sitting position. Slow, helpless groans jerked out of her. Her eyes rolled with wet whiteness, like the eyes of a hog in a slaughtering pen.

The eyes of a hog.[17]

In the 1976 film adaptation of the book, poor Carrie, played by the young Sissy Spacek, screams and emerges from the shower mist, throwing up her blood-covered hands towards her classmates. They delight in her bewilderment. 'Plug it up,' they jeer, throwing sanitary towels and tampons at her. A tampon gets stuck in her pubic hair. It's a sorry scene. Carrie is made to feel ashamed of her naivety. Pathetic. King's descriptions, painting her either as a startled farm animal ('She looked around bovinely', 'flailing her arms and grunting and gobbling') or quite unattractive ('fat forearms crossing her face'), do not help her cause. Rereading passages of *Carrie* for the purposes of this book, I felt sad at how many times King calls her dumb. Critics have long debated the misogyny of *Carrie* and the horror genre in general, but for me *Carrie* has always stood out because of the period scene in particular. It makes my chest hurt. Sure, she enacts her bloody revenge, killing men, women and kids with whole-sale ruthlessness, but the whole premise is that she's punished for *being* a woman. King might have explored the very real culture of shame that surrounds the female body in society, at a time when second-wave feminism was trying to emancipate the female body from centuries of oppression, but Carrie's menstruation remains a blight. She is eventually destroyed by the body she is unable to control, playing into the trope of female bodies being frightening, unstable and mysterious. Controlled by the moon.

Have we shaken off this notion of mystery? The idea that there is a hysterical monster lurking inside all bleeding women?

The happiness fantasy

Before looking more closely at how our mood can be affected by hormonal fluctuations, it's important, I think, to acknowledge the background pressure we all feel in our society to be happy and stable all the time. Having this knowledge in the back of our minds can help us understand what may be affecting our thinking when our mood dips, or when we feel very anxious, at certain points in our cycle.

The word 'happy' is tattooed on our psyche when we're small. When we're learning to understand our emotions with painted stickman faces, we take on the idea that happiness isn't just something that's nice to have, but something we *should* have. But for all the ways we obsess over happiness, it's tricky to say what it actually . . . *is*. Happiness is difficult to measure or define. God knows we've pondered it long enough. Aristotle believed happiness was tied up with how we exercise virtue and, rather than moments, was measured in lifetimes: how well we *lived*, not live. He maintained – cheerily – that few 'succeed' in being happy because it takes so much bloody work and discipline. Today, happiness is positioned not just as a right but a skill we can develop,

42

but in the Middle Ages happiness was long believed to be down to fate. The Old English root from which 'happiness' derives is *hap*, meaning chance. In the seventeenth century, views changed again. The revolutionary philosopher John Locke said the 'business of man is to be happy'[18]; that we should work to maximise pleasure and minimise pain. The idea stuck. If only we put our minds to it, grit our teeth and clench our knuckles white, we can, should, *will* be happy. But the happiness-as-a-natural-state idea has a big, throbbing sting in its tail.

If we see being happy as 'natural' it's easy, so horribly, seductively easy, to think there's something wrong with us when we're not. The torture lies in how we fight with our idealised versions of ourselves; how we *should* feel. We *are* moving away from the idea that if we aren't happy all the time it's our own fault, but we still rarely talk about issues like depression or anxiety without mentioning happiness in the next breath. I couldn't even begin to count the number of times that I have been having a rough patch, particularly when I'm premenstrual, and have said to myself: I just want to feel happy. As if I never have! As if I don't know what happiness feels like! I lose all perspective in those moments. Then the mood can slide away without me really noticing and I find that I'm not tight-chested, I'm revelling in the smell of meadowsweet in the marshes near my house, the crowns of fur in my dog's chest, making good dinners, making plans. If I am premenstrual, all this can happen in the space of an afternoon. Having a menstrual cycle sometimes feels like having access to a bonus level in the having-a-complex-human-brain game. But *all* moods come and go. No mental state is fixed, even if it feels like it's going to be.

Neuroscientist Dean Burnett unpicks the happiness-as-default thing in his book *The Happy Brain: The Science of Where Happiness Comes From and Why*. I had a chat with him over email. 'The idea is both unhelpful and harmful; a cultural invention caused by capitalism, competitiveness, pop-psychology, new-age influences and aspirational media. It seriously overlooks the crucial facts that neither the brain nor our lives are static; they're constantly changing.'

Headlines promising to reveal 'the key to happiness' (an exercise: count how many times you see those words over the next fortnight) are delicious because they tap into what we all feel: frustration at the lack of easy fixes for our pain. Solange Knowles evokes this torment beautifully in her song 'Cranes in the Sky'.[19]

If greater peace can be found in trying to accept that our minds are as fluid as the notion of happiness – in reality, an umbrella term for several processes happening in the brain when we experience something rewarding or enjoyable – I increasingly believe that this involves sometimes taking a step back from the focus we place on symptoms and labels (these serve different purposes for different people) in order to practise being with our discomfort rather than immediately trying to chase it away. Obviously, in the low moods, crying fits and heightened anxiety that can come with hormonal fluctuations, this may not always be possible. But in calmer moments, where we can, it's worth considering how we question ourselves in relation to what we believe happiness *should* feel like or where it *should* come from, and whether we can soften that line of questioning. Lasting happiness is also a logically flawed idea. As Burnett says, 'If you're happy anyway, why do anything? Permanent happiness would be, for the brain, like skipping to the end of a book to see what happens; you don't actually benefit from the knowledge as you would if you engaged with the story properly, even though you end up in the same place.'

Research increasingly emphasises how much happiness has to do with interactions with fellow humans – what our brains are geared towards. Most happy-making things we do (eating, laughing, travelling, sex) involve other people. Emotions like embarrassment and guilt only exist in relation to others. We know mental distress increases with loneliness and isolation. Many studies suggest that engaging in meaningful inter- actions with other people can make us report higher levels of happiness than material or financial rewards. That other people are so important to our happiness emphasises how the involvement of others when we're struggling is not just helpful, but crucial. How would it feel, then, to be a woman who mentally suffers each month and to have other women who know what that suffering looks and feels like to talk to? The number of Facebook groups and online forums in this area suggests that it's a very helpful thing indeed.

Seeking happiness and stability is one of the most potent motivators we have in this life, but we're doing ourselves a favour if we acknowledge that, like the tide, happiness comes in and out. The tidelike rhythm of our menstrual cycle and hormonal fluctuations can have a significant impact on our state of mind.

The cycle: a vital sign

Our menstrual cycle is like having another vital sign, like our heart-rate or blood pressure. We might not think of it as such, but the way our cycle 'behaves' can indicate whether everything is working as it should, whether there's something wrong or what can happen when we're going through a big change of some kind.

The cycle length is the number of days between periods – counting from the first day of bleeding until the day before the next period begins. The length of a woman's typical cycle is determined by all kinds of factors, including our genes, age, overall health, body mass index (BMI) and whether or not we're using hormonal birth control. The combined pill, mini-pill, patch, vaginal ring, injection or an IUD (coil – the Mirena, for example) are all hormone-based and may significantly affect the length of the cycle because they take over the regulation of our production of oestrogen and pro-gesterone. They halt the body's normal monthly fluctuations of these hormones by preventing the ovaries from maturing and releasing eggs to be fertilised.

Women not taking any form of hormonal birth control can expect to have a cycle length of anywhere between 21 and 45 days in the first two years of menstruating. Thereafter the average cycle length is 28 days, but it varies from woman to woman. In a sense, we are our own experimental control. Regular cycles that are shorter or longer than this, from 21 to 40 days, are considered normal, but one of the key indicators of reproductive health is the regularity part.[20] That said, it is normal to have *some* degree of variation in cycle length. For example, one cycle might be 26 days long and the next 32 days long. Cycle length and variation can be affected by stress, diets high in sugar or caffeine, jet lag, smoking, drinking alcohol and over-exercising, as well as by the use of hormonal birth control. Irregular periods are common after giving birth and, very often, women won't have any at all while they are breastfeeding.

Although our cycle is something essential we have in order

to procreate and, therefore, is a reasonably robust process, it clearly can be sensitive to some lifestyle factors. I am interested in how tricky it can be to *trust* a bodily rhythm that is, by its very nature, a bit variable.

No matter how many times I have read or been told that some cycle variation is normal, as soon as I happen upon a potential 'explanation' for irregularity, however minor, I wonder if it could apply to me. I have a propensity for anxious thinking and catastrophising, but I think, on some level, women often do, in relation to our mind and body processes. Particularly when you can't set your watch by them.

In my research for this book, I came across an American study from 2002 that looked at whether chlorination by-products in drinking water could affect cycle function and briefly scaremongered myself with it. This study found a pattern of reduced cycle length with increasing exposure to trihalomethanes (THM) (the by-product of chlorine being used to disinfect raw water supplies), suggesting THM might affect ovarian function.[21] I started reading more literature about THM and began to feel uneasy. I barely think about the water that comes out of my kitchen tap. I just drink it. I assume, I suppose, that someone has taken care of its safety along the way. There was a period, though, in the early to mid-noughties when everyone was obsessed with THM and water safety, even making links with increased risk of breast cancer. I had to stop reading, telling myself: 1) It is the twenty-first century and I am living in the Western world. I am lucky that the molecular make-up of my water supply is regulated and I can *probably* trust its safety. 2) Almost anything you could think of that might affect how the human body works has been the topic of a small-scale study. Still, I couldn't help but put my mind at rest by checking the sample size of this study (small – phew) and whether the conclusion said more research was needed to make any robust claims (it did – phew). I also looked at more recent studies of the dangers of THM for women's health and found them much more equivocal. A study by Spanish researchers published in March 2018 found that, at the levels of THM commonly encountered in Europe, there was *no* correlation between THM exposure and female breast cancer.[22] This satisfied me. My unease was quelled. Jesus.

If and when my own cycle fluctuates it tends to be shorter than I'm expecting. My five-minute meltdown with the THM business made no lasting imprint, but I did, for a second, look at the kitchen sink and quietly ask myself: could ... tap water be to blame? Should I not be so wildly *cavalier* with these everyday things I put into my body?

It felt significant to me, this little thought loop. Because when it comes to our health, even though we might be able to rationally understand that we are variable *by our very nature*, it can be hard to sit with the idea. We are, perhaps, due to centuries of societal fear around so-called female unpredictability, conditioned to be wary of it within ourselves. For some of us, this might mean we're susceptible to embracing neat-seeming explanations that pop up. I've no doubt there are women who would disagree with this idea and say that *surely* we're much stronger and more resilient than that. Some women may well be, but I, for one, have struggled to accept that there cannot be an explanation for every bodily or mental state. Even though I have known its rhythm for nearly twenty years, this is particularly true of my menstrual cycle.

At certain times of the month I find it harder to trust my body and mind, to accept that life as a woman can be messy and unpredictable. I know this is true for many women.

Perhaps we should be paying more attention to how our sex hormones can affect us psychologically. It's an interesting dichotomy, I think, that we can be quite disconnected from what's actually going on when our hormones are fluctuating, and yet be so ready to self-police when we feel anything other than sanguine and completely pain-free.

It is very easy to interpret our fluctuations in mood as inherently 'bad' or 'unhealthy'. We often follow moments of being 'hormonal' with increased self-surveillance. If you are someone whose mood can dip considerably before your period comes, have you ever said to yourself: 'I wish I wasn't like this. *Why am I like this?*' before slipping into an existential quagmire? Lord knows I have. It's a hard place. Obviously, the menstrual cycle is not the only process for women that involves hormonal fluctuations: hormonal birth control, fertility treatment, pregnancy, giving birth, miscarriage, breastfeeding, menopause and reproductive disorders like endometriosis and polycystic ovary syndrome (PCOS) all affect our balance of hormones, with all manner of effects on our physical and mental health. Understanding which hormones are doing what and when throughout the menstrual cycle, and to what effect, is the skeleton key to grasping the rest.

Each menstrual cycle can be divided into four main phases: menstrual phase, follicular phase, ovulation phase and luteal phase. In the description that follows here we'll work to a 28-day average cycle length and assume the period lasts five days (including that annoying, dregs-y bit at the end, when you don't know whether to bother putting a panty-liner in). If your cycle is shorter or longer than this the day ranges will of course be different, but you will get a feel for the structure of things and, if you do track your cycle, be able to apply it to yourself.

Menstrual phase (days 1 to 5)

The first day of your period is the first day of the cycle. A rapid drop in oestrogen and progesterone starts contractions in the smooth lining of the womb and this lining is shed through the vagina. The decrease in these hormones may trigger cravings for carbohydrate-rich foods like sugar (chocolate, cakes, biscuits, all the good things), bread and pasta. This is because a drop in oestrogen level can trigger a drop in the level of serotonin (one of our mood-moderators) in the brain, and our body tells us to eat more carbohydrates in order to help replenish it.

No two women will have exactly the same period experience. Some women don't bleed very much, others bleed a lot. The medical term for heavy periods is 'menorrhagia'. It can be accompanied by terrible pain ('dysmenorrhoea'). Heavy bleeding doesn't necessarily mean there's anything wrong with us, but it can have a profound physical and emotional effect, disrupting day-to-day life. The pain may be felt as cramps within the abdomen, spreading to the lower back and down the thighs. It can come in sharp spasms, enough to take your breath away, or be more constant. Gnawing. Pain can vary with each period, too. Some periods may be a breeze compared to others. Most months, I'll spend at least one day in bed or on the sofa in a co-codamol fug. Whenever I've had full-time jobs, I've always had to take an afternoon or morning off each month because of period pain.

This pain occurs when the muscular wall of the womb contracts (tightens). The womb is actually constantly having mild contractions, but they're usually so subtle we can't feel them. During the period, the womb's walls contract more intensely to encourage the lining to shed. This constricts the blood vessels that line the womb, temporarily cutting off the blood supply – and, consequently, the oxygen supply – to the womb. In the absence of oxygen, the womb tissue releases chemicals that trigger pain. Meanwhile, to help contractions, the womb lining is also producing compounds called prostaglandins – a key part in the period-pain story.

Prostaglandins play a key role in the generation of the body's inflammatory response: the immune system's way of tackling infection and injury.[23] Prostaglandins, found in almost every tissue in humans and other animals, are produced in the womb during the period because, really, there *is* an injury. The womb lining is breaking down and keeping it inside would not be good for us. Contractions are needed to expel it from the body. (This mechanism is the reason why prostaglandins are used in the induction of labour.) Studies suggest that women who experience more period pain may either have a build-up of prostaglandins or make more of them. Prostaglandins are also thought to cause the diarrhoea and excessive flatulence some women experience immediately before and during the period, by making the intestines contract along with the uterus. Lots of contracting means lots of, ahem, 'activity'.

As a general rule, more prostaglandins means more pain. Excess prostaglandins in the system is what is thought to be behind the muscular or rheumatic aches and pains, reminiscent of the symptoms of a cold or flu, that can make women feel as though they are 'coming down with something' just before their period. Women who have pain conditions like fibromyalgia, arthritis, endometriosis or migraines may find that they flare up either premenstrually or during the period itself because of all the extra prostaglandins being pumped out. As an example, a study of 1,697 women published in the *Journal of Headache and Pain* in 2015 found that nearly 60 per cent of women with migraine reported an association between migraine and menstruation.[24]

Severe period pain can sometimes be a sign of an underlying condition. Doctors refer to this as 'secondary dysmenorrhoea'. Conditions that can cause both severe pain and bleeding include:

- Endometriosis – cells that normally line the womb begin growing in other places, such as in the fallopian tubes, ovaries and bowel. These cells can cause intense pain when they shed. Scar tissue can start binding various structures to one another in bands called adhesions, causing digestive pain as well as pain when going to the toilet or during sex. In severe cases adhesions may cause bowel obstructions, which sometimes need to be operated on.
- Fibroids – non-cancerous tumours that grow in the womb
- Pelvic inflammatory disease (PID) – the womb, fallopian tubes and ovaries are infected with bacteria and become severely inflamed.
- Adenomyosis – tissue that normally lines the womb starts to grow within the muscular womb wall.

An intrauterine device (IUD) such as the Mirena or copper coil can sometimes cause bad period pain in the first few months after it's put in.

If heavy and painful periods are caused by an underlying condition, you may also have irregular periods, pain during sex, bleeding between periods or a less-than-rosy-smelling vaginal discharge. Any of these symptoms warrants a visit to your GP.

Follicular phase (days 1 to 13)

This phase also begins on the first day of the period. It carries on until halfway through the cycle. During this phase the hypothalamus stimulates the pituitary gland to release follicle-stimulating hormone (FSH), just as it did before our first ever period. The ovaries are made up of sacs (follicles) and the eggs live inside these, suspended in fluid. Follicles are at different stages of development in the ovaries at any given time. FSH does exactly what its name suggests and causes one follicle in one of the ovaries (known as the 'dominant follicle'), along with the egg it contains, to mature and grow. It can grow to be up to three centimetres – about the size of a strawberry.

As the follicle grows (towards the latter end of this phase) it produces a surge of oestrogen. At this stage, the follicle is the main source of oestrogen in the body. A change in our vaginal discharge usually happens; it becomes wet and slippery, like the consistency of raw egg white.

When oestrogen reaches its threshold level, the tiny little egg is ready to begin its journey. After oestrogen has peaked, the brain produces a burst of luteinising hormone (LH) to trigger ovulation. Around twenty-four hours later, using enzymes to break down its own wall and create an opening, the follicle makes way for the egg to be released.

This major spike of oestrogen just before ovulation can, depending on your individual sensitivity to hormonal fluctuations, cause headaches, bloating, tiredness, nausea, low mood, anxiety and trouble sleeping. For years, I would have these episodes of bilious, skull-thrumming anxiety that seemingly came out of the blue, when there was no major stressor or familiar trigger (you do get a sense for them as the years go by). I'd get nervous about what was happening to me, thinking: *Why why why?! There's no reason for this!* This secondary layer of interrogation, the frantic, explanation-seeking thoughts, sends the pressure valve right up. Then, what was an afternoon of feeling a bit squiffy can turn into days on end being sure of a pending breakdown.

After downloading a period-tracking app and putting in my data, I realised that these 'out of the blue' episodes seemed to happen most before ovulation. It is no exaggeration to say this has been a revelation for me. After an entire adult life spent negotiating with anxiety of varying colour and shape, along with studying the human mind, I know full well that the notion of there ever being an *exact* reason for a period of low mood or anxiety is an anathema. We are always reacting and adapting to the squally waves of life. Stressors like demanding jobs, relationship issues, money worries and illness do, funnily enough, cause stress, and our individual resilience – or vulnerability – is the product of many complex variables in our nature and nurture. But, *but* – knowing that my biology is *part* of the picture, that a hormonal surge is a temporary thing, sometimes gives me a touchstone. When I look at my calendar I know I *could* feel a bit strange on those days. I also know that it will pass. This, over time, has made me feel a bit more equipped.

The anxiety circuit board

We know that female sex hormones can have an effect on the brain, affecting our mood and stress responses. This is thought to be because of their effect on several neuro-transmitters, the brain's chemical messengers. Research has not yet established *exactly* what is going on in our bodies when we become highly anxious and uncomfort-able, because the underlying mechanisms are extremely

▶

---

complicated. However, as we continue to explore the mind–body connection it may be helpful to know what has been discovered to date.

The amygdala, a little almond-shaped section of nervous tissue that is central to our emotional processing, interprets a potential danger (this can be our own anxious thoughts) and sends an alarm signal to the hypothalamus, which communicates with the rest of the body through the nervous system so we have the energy to either fight, freeze or run away (flight). The hypothalamus activates the sympathetic nervous system, which is continually active at a very basic level to maintain homeostasis (the maintenance of internal stability within the body, such as temperature) but has the primary role of stimulating the fight-or-flight response. Signals are sent to the adrenal glands, which respond by pumping out the hormone epinephrine (adrenaline) into the bloodstream.

As adrenaline circulates through the body, physiological changes happen. The manifestations of anxiety experienced will probably differ from one person to the next – there are literally thousands of reported symptoms from across the world – but we know that the heart may beat faster than usual, thrusting blood to our muscles and vital organs. Blood pressure may go up. We may start to breathe more rapidly, and small airways in the lungs expand so we can get as much oxygen as possible with each breath. All the extra oxygen is sent up to the brain to increase alertness and sharpness of senses. The physiological manifestations of anxiety for me usually include nausea, a stinking headache, pins-and-needles, dizziness and the feeling that my bowels could become a weapon of war. A laugh riot, in other words. It's hard to believe when it's happening, but there is an evolutionary function behind the bowel chaos. When our brain interprets that we are under attack, our blood flow shifts from the intestines to the heart, lungs and muscles so we can run faster. It's a primal reaction that hasn't quite caught up with the modern business of being a human; in prehistoric times, this diversion of blood flow away from the gut would prompt us to empty our bladder, bowels and stomach, to shed literally any gram of matter weighing us down in order to run from bears, smilodons and mammoths at the speed of light.

While all this is happening, adrenaline is also triggering the release of glucose (blood sugar) into the bloodstream to supply energy all over the body. This entire system activates so quickly and imperceptibly that, sometimes, we're not even aware of it. In fact, it's so efficient that our brain can start this hormonal fairground ride before our visual centres have a chance to process the incoming information and we consciously see the danger. This is why we're able to do things like dodge a speeding cyclist in our path before we've even *thought* about what we're doing, giving us that *shit, that was close* feeling.

After the big surge of adrenaline starts to drop, the hypothalamus initiates the second part of the stress response system: the HPA axis. (Whenever I see it written down I imagine Alan Partridge saying it: quite a good learning cue, it turned out, for the psychobiology exam that

▶

was part of my Psychology MSc course.) This network contains the hypothalamus, the pituitary gland and the adrenal glands. If the brain continues to perceive something as dangerous, the HPA responds to a series of hormonal signals to keep the sympathetic nervous system's accelerator held down. The hypothalamus will now send out corticotropin-releasing hormone (CRH), which travels to the pituitary gland. This triggers the release of yet another hormone, adrenocorticotropic hormone (ACTH). This travels to the adrenal glands and stimulates the production of cortisol, which is a steroid hormone like oestrogen and progesterone.

As long as cortisol is at the party, the body remains on high alert. Cortisol levels drop when the threat – whatever our mind perceives that to be – passes. The parasympathetic nervous system, the 'brake' for all this commotion, calms down the stress response. Many people find putting on that brake quite difficult. I'm one of them.

Regular stress keeps the HPA axis activated. Interestingly, one of those stressors can be going without food for too long. When we're hungry, there is an increased release of a hormone called ghrelin from the stomach, which heightens our motivation to seek and consume food. Studies have shown that raised levels of ghrelin might also activate the HPA axis. There are receptors for ghrelin in the hypothalamus and, the more this hunger hormone circulates, the more anxious we feel. This might be where you would say you feel 'hangry': a word now defined in the Collins Dictionary as an adjective, meaning 'irritable as a

▶

result of feeling hungry'. There have been animal studies showing that injections of ghrelin direct into the hypothalamus increase anxiety-like behaviour, suggesting there are overlapping brain circuits mediating food intake and emotional behaviour. The gut and brain, then, essentially have a hormonal M1 running between them.

Fortunately there are ways of dealing with the effects of the HPA axis being activated – not least having a sandwich before you get so hungry your hands are shaking and you wage a quiet war on everyone around you. There is a huge body of evidence now that tells us regular exercise will decrease the amounts of cortisol and adrenaline in the bloodstream. Regularly practising relaxation techniques such as mindfulness can lead to enduring changes in brain function, stopping the amygdala from sending alarm signals so readily. The basic functions of our being deserve respect, too. I know, I know, it's boring to read this. But we sort of *want* our bodies to be boringly effective because when they're not, we can feel bad. We are organic machines that require certain things if we are to function optimally. The fundamental formula of human life is, ultimately, that we suffer in some way when we don't give ourselves what we need: a varied diet, enough sleep, enough water and regular connections with other human beings.

Ovulation phase (day 14)

Ovulation usually happens on around day 14 of the cycle. In a 28-day cycle with four days of bleeding, this would be roughly ten days after bleeding has finished, and two weeks before the start of the next period. This is an incredibly rough guide, though, as no two cycles are ever identical.

Around twenty-four hours after the brain has triggered the big surge in luteinising hormone (LH), the egg ruptures its sac. This is ovulation. Although the egg seems very small – around 0.1 mm in diameter, about the size of a grain of sand – no other cell in the body is as big, or visible to the naked eye. Under a microscope it looks like a planet; fitting for something with the power to start a new human life.

While some women feel nothing when they ovulate, others, including me, find it quite painful. The pain of ovulation, known as *Mittelschmerz* (a snappy portmanteau of the German words for 'middle' and 'pain'), can be experienced as anything from a mild twinge that lasts a few minutes to significant discomfort that lasts a couple of days. Pain is generally felt on one side of the lower belly or pelvis; which side that is will vary from month to month depending on which ovary is releasing the egg. Nausea is common if pain is severe. I often have twenty-four hours of vague, sweaty greenness around this time.

When we ovulate, the fluid inside the follicle and a small amount of blood are released by the ovary. It is thought that either the fluid or the blood, or both, may sometimes travel through the ovary wall and irritate nerves in the lining of the abdominal cavity; however, the exact cause of ovulation pain is unknown.

To give you an insight into how painful ovulation can be for some women, I recall a conversation I had about ovulation with a consultant gynaecologist I will call Mr C. I was referred to his NHS clinic to discuss potential PMS treatments. We spoke about ovulation at length because around this time tends to be when my psychological symptoms are worst. Mr C told me that he quite often sees women pre-senting in Accident and Emergency in tears with acute lower abdominal and pelvic pain, whose only clinically significant finding turns out to be recent ovulation.

Endometriosis can cause significant pain during ovulation, as can scar tissue (for example, from a caesarean section or appendix removal) that restricts the ovaries and surrounding structures such as the bowel. STIs like chlamydia that can cause inflammation and scarring around the fallopian tubes may also lead to ovulation pain.

Luteal or 'secretory' phase (days 15 to 28)

This phase immediately follows ovulation and continues until the end of the cycle. During this phase the egg is moved along the fallopian tube by cilia – hair-like cells in the tube's lining. A ripe egg will 'live' in the fallopian tube for twelve to twenty-four hours. This is how long it has to get fertilised by sperm. As we're always warned, sperm can survive for quite a while longer than an egg – up to five days. This means that if sperm have entered the womb a few days before the egg finds its way into a fallopian tube they can still fertilise it.

Progesterone levels begin to rise now, in order to prepare the womb in anticipation of an egg being fertilised and implanting itself within the lining. Progesterone is a sedating hormone and, as levels increase, it can slow us down. We might feel physically tired, mentally foggy and introverted, desperate to live under a duvet in a bobbly tracksuit. An ex-partner of mine had a word for this way of feeling: *innich*. She would put her arms around herself when she said it. Apparently it's German but I couldn't find a translation. She also doesn't speak German, so who knows. I have nonetheless continued to use the word because it just fits.

The National Association for Premenstrual Syndrome (NAPS), an important resource for women suffering as well as for healthcare professionals not trained in this area of women's health, say: 'PMS is categorised by a number of symptoms – over 150 have been identified'.[25] They list the common psychological and behavioural symptoms as: mood swings, depression, tiredness, fatigue or lethargy, anxiety, feeling out of control, irritability, aggression, anger, sleep disorders and food cravings. Common physical symptoms are listed as: breast tenderness, bloating, weight gain, clumsiness and headaches. Their guidelines continue:

> No one experiences all identified symptoms. One symptom may be dominant. Each symptom may vary in severity during a cycle and from one cycle to another. New symptoms may present during a woman's experience of PMS. PMS symptoms may be experienced continuously from ovulation to menstruation, for 7 days before, at ovulation for 3–4 days and again just prior to menstruation and in other patterns. Some women do not experience relief from symptoms until the day of the heaviest flow.

There is compelling research to suggest that women who really suffer with premenstrual syndrome each month might be very sensitive to their fluctuating levels of progesterone. Like the other major sex steroid, oestrogen, progesterone can have a powerful effect on the body. It can cause constipation by slowing down digestion, water retention (including the delightful bloat that comes with it), and a big surge in appetite, along with cravings for calorific and fatty foods. A study by Tunisian researchers published in the journal *Annals of Endocrinology* in 2016 found that women can eat up to an extra 500 calories a day in the days before their period.[26] All this because our body thinks we might have got pregnant during ovulation and wants us to absorb more nutrients from the food we're eating. Progesterone literally wants us to eat enough for two. Interestingly, there is also evidence to suggest that high levels of progesterone in the body can slow down wound healing during this phase. A study published in the journal *Clinical and Experimental Dermatology* in 2015 found that, at peak levels of progesterone, conditions like acne, psoriasis, eczema and dermatitis could become worse.[27] This is thought to be because of reduced immune function, i.e. less ability to fight off disease. In the follicular phase, when progesterone levels are lower, the body appears to be able to heal much faster.

Eating too little during the luteal phase can leave us vulnerable to significant shifts in mood. This is because high levels of progesterone make us more sensitive to drops in blood sugar. It's a good idea to eat regularly and not let ourselves get too hungry. I carry snacks with me at all times, like tiny mood flotation devices. Nuts fill my pockets and spill across my desk. Bananas bruise and burst through their skins in my bag, making it smell like a baby's car seat. Although cravings for high-sugar foods – I crave cheap kids' birthday party cake, specifically Marks and Spencer's Colin the Caterpillar cake, every single month, without fail – might be hard to resist at this time and bring us instant pleasure when we indulge them, it's probably not a great idea to eat *too* much, because then the spikes in our blood sugar levels will be quite dramatic, quite often, causing even deeper troughs immediately afterwards. Given that we're especially sensitive to low blood sugar in this phase, those dips are likely to make us feel a bit shitty. High progesterone levels are thought to cause our libido to drop but make us feel emotionally closer to our mate. This might be why we crave a lot of cuddles, hand-holding and hair-stroking rather than hot sex.

Towards the end of this phase, the egg disintegrates if it doesn't become fertilised. This causes levels of oestrogen and progesterone to start decreasing rapidly. The drop in these hormones initiates contractions of the smooth lining of the womb (which can start a day or two before the period starts – hence the pain we may feel in the lead-up to bleeding) and the menstrual phase begins once again. At this point, many women who suffer with psychological symptoms of PMS report feeling a marked sense of relief. I quite often feel a sense of euphoria around this time, an almost druggy feeling of wellbeing, and discover when I go to the bathroom that my period has started. This pattern has happened too many times now to be coincidence.

The thyroid

A possible cause of irregular periods or heavy bleeding is a thyroid disorder. If you are really suffering each month and haven't ever been to your GP to have it investigated, it is likely they will want to check your thyroid. If they don't, ask them to. A simple blood test is all it takes.

The thyroid is a bow-tie-shaped endocrine gland found in the neck, in front of the larynx (voice box), made up of two lobes, each around the size of a halved plum and sitting either side of the windpipe. The thyroid makes two hormones that are secreted into the blood: thyroxine (T4) and triiodothyronine (T3). In our body's cells and tissues, T4 is converted to T3. It is the T3, derived from T4 or secreted as T3 from the thyroid gland, which is biologically active and influences the activity of every cell. This hormone is essential to keep cells functioning properly to maintain our metabolism (the chemical process our body uses to turn food into energy) as well as heart and digestive function, muscle control, brain development, bone maintenance and mood.

Both men and women can contract a thyroid disease, although it's more common in women, and it can have significant effects on reproductive health. Thyroid disease is divided into hyperthyroidism (overactive thyroid) and hypothyroidism (underactive thyroid), and there are numerous causes of these diseases. Another type of thyroid disorder is Hashimoto's disease, which is essentially a variant of hypothyroidism but caused specifically by autoimmunity. This means that the immune system starts attacking the thyroid gland; but it's not really understood why this happens. As the thyroid gets destroyed over time, sometimes beginning with an overactive phase, it is unable to produce enough thyroid hormone and becomes underactive.

The symptoms of an underactive thyroid include excessive tiredness, sensitivity to the cold, weight gain, depression, constipation, sluggish thinking, muscle aches and weakness, dry skin, brittle nails and hair, loss or thinning of the hair, and tingling sensations in the hands and fingers (carpal tunnel syndrome). An overactive thyroid can cause pretty much the reverse of all these symptoms: anxiety, irritability, trouble sleeping (causing tiredness in the day), sensitivity to heat, fast or irregular heart rate, trembling, weight loss and, sometimes, a swelling in the neck when the gland becomes enlarged (a goitre). Graves' disease, named after Robert Graves, the Irish doctor who began describing patients with the condition in the nineteenth century, is the most common cause of hyperthyroidism in the UK and is much more common in women than in men. Some patients with Graves' disease have a goitre and some also develop eye problems, known as thyroid eye disease. The eyes may become prominent, appearing to bulge out from the face, and be quite painful. There is an established link between Graves' disease, an autoimmune disease thought to be caused by the body's immune system mistakenly attacking the thyroid, making it become overactive, and endometriosis.[28]

Sometimes, an underactive thyroid can make periods heavier. Research suggests that this might be due to one of three mechanisms: without enough thyroid hormone the ovaries may not be able to make enough progesterone, which has flow-decreasing properties, so bleeding may be heavier; insufficient thyroid hormone may mean you are not making enough of the coagulation (clotting) enzymes you need to stop heavy bleeding; without sufficient thyroid hormone you are making less of the protein called sex hormone-binding globulin (SHBG), which binds tightly to oestrogen and limits how much of it our body is exposed to. Without enough SHBG we're exposed to more oestrogen. When oestrogen levels are very high, the lining of the womb can become very thick, causing heavy periods.

Impaired thyroid function can make some symptoms of PMS worse for women, too. If you are feeling tired, sluggish, achy and sad before your period because of fluctuations in sex hormone levels, it makes sense that an underlying thyroid disorder with many of the same symptoms might make you feel worse. As part of my Psychology MSc requirements I worked with a clinical psychologist in the NHS who oversaw two multidisciplinary teams in bariatrics (weight-loss surgery) and a chronic pain service. It was compelling how many women seemed to present with a triad of chronic pain (disorders like fibromyalgia) or fatigue, thyroid disorder and a reproductive condition such as PCOS, endometriosis, or, indeed, uncomfortable PMS symptoms. The reason for these connections could never be attributed to a single factor, although advocates of naturopathic medicine may plead otherwise.

Something called 'adrenal stress' or 'adrenal fatigue' is often cited in holistic medicine, particularly in regard to women's health. The last time I went to Los Angeles to visit my sister who lives there, it's all I heard anyone talking about. The theory goes that the adrenal glands (the endocrine glands lying just above the kidneys that produce a variety of hormones including adrenaline and cortisol) become 'fatigued' and less effective when the body is under a lot of stress, be that emotional or with chronic illness. The main symptom is said to be, well, fatigue, but 'cravings for salty foods' and caffeine dependence are also frequently listed. There is absolutely no robust scientific evidence that suggests 'adrenal fatigue' is a thing. Systematic reviews of existing research establish again and again that it does not exist. However, there is money to be made from pushing naturopathic supplements and 'healing' treatments to women who are at sea in how they feel, perhaps experiencing symptoms of a different underlying organic problem, or just feeling exhausted by life. The placebo effect of sitting in front of someone who says they know exactly what is wrong with you, who listens carefully, can offer very scientific-sounding explanations and, in some cases, tests, can be very strong.

The truth is that the way our various hormone systems interact with brain function, and the impact stress and anxiety can have on the way our hormones work, are very difficult to measure precisely. It is known that thyroid function can change during pregnancy due to the higher levels of oestrogen, but the more general interaction is a complex business. It makes sense that if a woman is wrung out by the symptoms of something like endometriosis, her general mood levels would be affected. Pain is tiring. Being tired all the time can make one feel sad and apathetic.

A test measuring your hormone levels is the only accurate way to find out whether there's a problem with your thyroid. A blood test called a thyroid function test looks at levels of thyroid-stimulating hormone (TSH) and thyroxine (T4) in the blood. A high level of TSH and a low level of T4 in the blood could mean you have an underactive thyroid. If your test results show raised TSH but normal T4, you may be at risk of developing an underactive thyroid in the future and your GP might recommend that you have a repeat blood test every so often to monitor the situation.

The majority of women with PMS issues will not have a problem with their thyroid. However, if symptoms are really getting in the way of your day-to-day functioning, it is always worth telling your doctor. If a thyroid disorder is detected it can be treated very effectively. An underactive thyroid is treated by taking daily tablets of a hormone replacement called levothyroxine, which replaces the thyroxine hormone your thyroid isn't making enough of. You'll initially have regular blood tests until the correct dose of levothyroxine is reached. In most cases you'll have to take the tablets for the rest of your life. Anti-thyroid medications are used for the treatment of an overactive thyroid, that prevent the thyroid from producing excess amounts of hormones.

Part Three

Varium et mutabile semper femina.
(*Woman is ever a fickle and changeable thing.*)

– Virgil, *The Aeneid*

Part Three

The female body has been positioned as something that needs to be regulated and controlled for centuries. Still to the present day, when we express pain that relates to our body in some way, what we say or do is ripe for misinterpretation or dismissal. Other people – usually men with more power – have historically decided on the meaning and worth of our pain. This can make it difficult to talk about what causes us distress because we fear how we'll be perceived. Time and time again we are given the message that something else, someone else, is worth more than our discomfort. There is a clear link between the historical treatment of 'hysterical' women – *any* woman, that is; inherently weedy but also mercurial and untrustworthy, acting under the influence of her mind-controlling hormones – and the #MeToo movement, for example, because so, so many women were told that their silence had more currency than their trauma. 'Be quiet and we'll reward you' was the message. Women were led to believe their future successes were predicated on packaging up their pain and putting it somewhere where no one would hear or see it. Traumas were frozen in time. Male reputations were protected above all else.

Women still, in almost every area of society, find it hard to get their pain taken seriously, and I think it's important we examine why. Sometimes, we don't even take our own pain seriously. We have been conditioned to view anything other than sanguinity as departing from the ideal because, historically, we've been punished for our so-called excesses. It took me years to admit in a clear voice, to myself or anyone else, that I sometimes have a really hard time with premenstrual symptoms: the gill-greying nausea, the headaches, the backaches, the drum-tight breasts, the acute anxiety and bouts of low mood characterised by self-criticism, the tears. I say 'admit' because, as a woman, verbalising pain that is indelibly linked with *being* a woman is a loaded thing. These symptoms pass, but they can be exhausting. I get on with day-to-day life for the most part, but the experience really does something to my sense of self some months. Even in the conversations we have with ourselves *about* ourselves, we carry history in the language we use. It's in our skin, muscles, bones, pink and grey matter. As Jane Ussher, a psychologist who has published extensively on our constructions of the female body and women's 'madness' over the last thirty years, writes in her book *Managing the Monstrous Feminine*:

> Throughout history, and across cultures, the reproductive body of woman has provoked fascination and fear. It is a body deemed dangerous and defiled, the myth of the monstrous feminine made flesh, yet also a body which provokes adoration and desire, enthralment with the mysteries within.[29]

72

Ussher and other critical psychologists say we must take into account how subjective the female body is and, therefore, how subjective the experiences the woman who inhabits that body will be. Perhaps it is this history of myth and fear Ussher speaks of that made me feel like some sort of traitor for admitting that PMS knocks me sideways. After all, women have been fighting so long to shake our image as hysterical, chocolate-hogging slaves to the moon cycle. Why would I want to admit that I suffer or am somehow less capable, stable or emotionally reliable by virtue of *being* a woman?

It has to be the truth, though, that my fluctuating hormones are part of the tapestry of my anxiety in life. Stressors like heavy skies, financial worries, arguments and having to take the dog to the vet for the third time in a fortnight vary in how manageable they feel. This shift in my resilience is the product of many things, but my biochemistry has to be part of it, right? I wonder what would happen if I wasn't who I am: a tenacious, educated white woman who has the time and space to really consider my mental health. Maybe, without the forensic focus I place on myself as a woman in the Western world, I'd know greater peace – or be less inclined to report otherwise. How do I decide when to stop seeking this hallowed hormonal 'balance' people talk about and start accepting myself as a walking process, a being in constant movement and change? Is that something I can possibly do?

Pathological femaleness

It is common sense rather than the reserve of feminist critical theory to suggest that society's overarching views of the processes of womanhood will affect our own perspective. We have to think about where feminism itself fits in, too. My fear of pathologising myself was a big part of why I resisted any kind of treatment for PMS for so long. *It cannot all be because of my womb.* But at the same time, my hormones appear to have the power to hijack my 'normal' state of mind. This creates a lot of inner tension.

What should I settle for? Where do I draw the line at what is or isn't my 'lot' and try and get on with it? I went on a journey of acceptance with my anxiety, acknowledging the complexity of traumatic things I experienced in my early life and, with good therapists, reached a place of knowing that my propensity for anxiety will probably always be there, but that I can manage it. Most of the time. Should I be applying this way of thinking to what is clearly a trigger of that anxiety: my hormones? Why do the treatment options for women like me whose mental health appears to be significantly impacted by our hormones seem so limited? Is the picture going to improve in the future? Could it be that our public health-care systems have not yet caught up with the lived realities of being a woman and that treatment only becomes joined-up, nuanced and individualised when we can afford to pay through the nose for it?

In order to answer these questions, we have to acknowledge the basic fact that the stigma surrounding female reproductive processes is still alive and well. I know this, we all know this, because otherwise we would not feel that crack of relief when someone dares to start talking about periods, the reality of birth, miscarriage or menopause, and other women say, *Oh thank god I can talk about this.*

This stigma is like an iceberg. The tip above the water's surface is women's day-to-day suffering with their menstrual cycle, pregnancy, menopause or whatever process they're going through; a suffering they feel unable to fully express, perhaps through fear of embarrassment, fear of disgusting others or the worry that they'll be dismissed. The deeper structure of the berg is the way women's health and pain has been, still is, viewed and treated in society, and all the myriad problems that come with that.

When I was on the Tube in rush hour one evening last summer, I looked around at all the women in the carriage. Middle-aged women in suits hurtling towards the suburbs after work. Pregnant women holding their lower backs as they got up off their seats. Teenage girls slick with sebum and Marc Jacobs Daisy. Elderly women standing stoically by the doors. Every one of them will have some story about how their reproductive systems have affected their lives. I wondered who they have spoken to about it, how much pain they've known. I looked at one of the older ladies resting her head on the glass and thought: how much have you kept a secret?

Imprinting

The world outside the skin influences the one within it. The body does not exist in a sociocultural void. The membrane between us and the world is permeable in more than one sense. We are not just our flesh. Of course, we are born into this world with a genetic make-up that will predetermine certain physiological characteristics (hair, eye and skin colour) and, in some cases, pathologies (disease), but how much does the world and the society we're born into shape the way we think?

The nature versus nurture debate is one of the oldest philosophical issues within psychology, if not *the* issue of how we understand ourselves as a species. It is generally agreed today that both factors play a critical role, but it's worth revisiting some of the historical ideas in the roots of the field, particularly if we're concerned with the connection of body and mind.

Hopping back to the seventeenth century and John Locke's *Essay Concerning Human Understanding*: he popularised Aristotle's *tabula rasa* theory ('scraped tablet' in Latin, but usually translated as 'blank slate') that the human mind is born 'blank', with no innate ideas. By extension, our minds are formed by experience and knowledge – imprinted on the slate – alone. Although Locke's position has been contested several times over the centuries, and is at odds with the findings of modern neuroscience. As our tools for examining the brain have become more sophisticated, we've learned that it is, in fact, pre-organised and programmed to process sensory input, emotions and motor control; but we also know that the parts of the brain dedicated to these processes refine their performance over time. However, Locke is still considered to be one of the biggest influences on epistemology: the study of the origin, nature, methods and limits of human knowledge. The idea of a 'blank slate' may be attractive to some who find the notion of being determined by their genes disquieting. On the other hand, the theory also implies there are no limits to how society can shape a human being's psychology. Locke's position, even given our awareness of how complex the interaction of different variables is in shaping what we know and how we get along in this world, speaks of a truth we all know: just how porous we are to the world we're born into. What and who is around us at different stages of our lives. This idea of imprinting is important when it comes to understanding women's mind–body relationships.

Right from when I had my first period, I felt the need to hide my fertile body. The messy parts of it, anyway. The allure of my breasts, legs and other plains of bare skin became clear as I got older. I have both celebrated and rejected that allure at various points in my life. The instinctive embarrassment and shame surrounding the natural process of menstruation, however, has always *been there*. It came with the blood and the pain; a message from the flesh. For so many years this shame and, later, the bewilderment I felt towards my own body, was unexpressed. The words never left my mouth because I didn't know which ones to choose. How do we begin to explain such a thing? Render speakable the unspeakable?

Although it might be depressing to hear in modern Western society, where women are fighting harder than ever for systemic equality, it is a fact that we have been told for centuries that our reproductive bodies are freakish and need controlling. Regulating. Society moves on, but slowly. As a woman born in the 1980s and now in my thirties, I am a product both of the progressive society I inhabit now and the edicts of societies that came before me. History does not go away. We can only really get to grips with why stigma surrounding our reproductive processes still exists if we spend a little time considering what's come before.

In *Managing the Monstrous Feminine*, Jane Ussher argues that menarche signals the beginning of the surveillance of the fertile body, marking the point at which 'a girl becomes a woman; when childhood innocence may be swapped for the mantle of monstrosity associated with abject fecundity'. It is absolutely true that, across cultures, our bodies are positioned differently once we've started bleeding. The blood often becomes a sign of pollution, needing careful disguise. Menstruation itself is seen as a cause of weakness and unreliability. Yet we also begin to be seen as sexual objects, both seductive and threatening: a complicated role we play for the rest of our lives until menopause comes around and the absence of the menstrual cycle – along with our ability to bear children – marks us as dry, undesirable old hags. A whole new ecosystem of stigma spins into life.

There are, of course, women who have a positive and fun experience of their first period. I have a couple of friends whose mothers threw them little parties. It seems fair to suggest that, if the beginning of menstruation was a positive affair, it was probably experienced by someone who was prepared. Someone for whom bodies and all they do were part of day-to-day conversation. Obviously, our parents influence the relationship we have with our bodies a great deal. There are so many different cultural norms attached to our bodily emissions, and the ways in which we censor or shy away from them are likely down to what we are taught and absorb throughout our early lives.

Leaky

I always find people's different hierarchies of which bodily functions are and are not for public consumption – for want of a better phrase – fascinating. Most of us have no problem loudly blowing mucus from our noses into a tissue in a public place if we have a cold. Nor do we mind peeing loudly in a public toilet stall within earshot of others doing the same thing. Men do it standing next to each other, unpartitioned, at urinals. Some people have no issue with pooing in public toilets; little time for faffing around with paper-down-the-bowl or expertly-timed-flush trickery to cover the sound. (There are toilets in Japan that do all this for you and more. I so long to know how it feels to have 78 sequential variations of water pressure directed at your naked buttocks while gagaku music seeps gently from a speaker.) Others cannot bear the horror of someone else hearing them do a (activating my father's voice here) 'big job'. Some people see no problem with spitting in the street or burping on public transport. The reality is that, while we're all smelly, noisy and leaky beings, we all have different psychological boundaries concerning what we release, where, and how we talk about it.

As a child, I was part of a very naked household. Everyone piled into the bath together, went to the toilet with the door open and broke wind with abandon. My siblings and I must have seen naked flesh as little more interesting than the carpets we ran around on. My sister and I groomed each other like chimps and, when we can, given that she lives on the other side of the world to me, still do. We were all taught about manners and about how what you do or talk about at home isn't always what you do or talk about in public or around people you don't know very well. As an adult I have rarely been squeamish in discussing what my body does. I'm able to read my audience, obviously, but my baseline is most definitely 'sharing'. I have, by and large, always had what feels like a healthy relationship with my functions. However, the total openness I had as a child completely changed at the junction of puberty.

As my body developed, I became swoony with embarrassment when my mum hung out my Tammy Girl training bras on the washing line. When I got my first period the idea that my bloody underwear or even the box my tampons came in would be seen by anyone – even my sweet, understanding and apparently unfazed dad – was just terrible. These functions were, are, different to the rest. Why? Why does blood leaving a woman's body, my own body, have such an edge in the squeamishness stakes? Not just the blood itself, either, but its *meaning*: woman, temperamental birthing machine.

I'm reminded of how much international attention Kiran Gandhi, a 2015 London Marathon runner, got for free-bleeding throughout the race. In her final hours of preparation, Gandhi got her period. 'No man I know would put cotton between his balls and run twenty-six miles,' she said on the ABC podcast *Ladies, We Need To Talk*.[30] 'In a radical act to prioritise my own comfort, I decided to bleed freely and run.' She knew it would be a radical move. 'I knew it was combating stigma and my own shame in my own right.' Google Images is full of pictures of Gandhi completing her first marathon with a big dark stain between her legs. Even as a woman who is fascinated by her own emissions, I find it quite startling to see because we just *don't* see that. We hide it at all costs. 'It was so empowering,' she said. 'I was like, "Damn! I'm running and bleeding" . . . I felt an enormous sense of power.'

Gandhi's free-bleeding was particularly powerful because we know that women often go to great lengths to conceal their menstrual blood. Dr Carla Pascoe, a research fellow at the University of Melbourne, has been studying how attitudes around menstruation have changed over the last century. In an interview with ABC Australia, she said the taboo has become 'subtle and complex'. 'The major way we can tell there is still a menstruation taboo is that you can still make money from it. If you analyse the advertisements from sanitary product companies, most of them are "buy our product because we can offer you a more effective way to conceal menstruation".' For a paper called *Silence and the History of Menstruation*, Pascoe interviewed women of different ages who revealed the lengths they would go to in order to hide their period.

Women told me if they go to someone's house and there is no bin in the bathroom, they would bundle up a pad or tampon into toilet paper and stick it in their bag and take it home. These are grown women. Even with adolescent girls, one girl was wrapping them up and putting them in a bag under her bed because she didn't want to be seen using the family bin.[31]

Pascoe said that women also hid their sanitary products while shopping at the supermarket and found it difficult to talk about their periods, even with their romantic partners. 'Often even within the spaces of private relationships there can be that discomfort,' Pascoe continued. 'Women find it difficult negotiating whether to have sex or not during their period . . . whether men will be grossed out by it.'

'Grossed out' is such an interesting term because it is usually involved in the exposure to something animal or visceral. It implies that, in disgust, a person is ejecting themselves – similar to the intriguing phrase 'beside myself'. We use that all the time when describing highly emotional experiences, positive or negative. The literature inevitably traces the phrase back to Ancient Greece, where 'beside' meant 'away from' or 'outside of'. To the Ancient Greeks, distressing or euphoric experiences could cause a person's soul to exit the body and actually stand beside them.

One of the finest voices on our fear of leaky bodies is the renowned Bulgarian psychoanalyst and philosopher Julia Kristeva. In 1980 she wrote an extended essay called *Powers of Horror* on the subject of abjection. According to Kristeva, the abject refers to the reaction of horror humans have to the threat of a breakdown in meaning. The essay opens thus: 'There looms, within abjection, one of those violent, dark revolts of being, directed against a threat that seems to emanate from an exorbitant outside or inside, ejected beyond the scope of the possible, the tolerable, the thinkable.'[32]

This reaction, Kristeva posits, is caused by the loss of distinction between the self and other; ourselves and something else – or, usually, some*one* else. What as a rule causes this reaction most intensely is the corpse: the most chilling reminder of our own organic nature. To see a human corpse is to be reminded that we are made of stuff that can, and will, die, decay and spore like a piece of rotting peach. These days, most of us probably won't see many corpses in our lifetime, but Kristeva believed that the same reaction can be triggered by other things. Namely, bodily emissions. Pus, sweat, breast milk, faeces, urine, semen, mucus. Blood.

These all signify a body without boundaries. Without boundaries, our sense of ourselves as controlled, contained beings is threatened. Seeing, hearing or smelling them, however briefly, can make us feel we are in and of each other. Amorphous. Unstable. To some primal part of us this is threatening. Revolting, in the truest sense. We can apply this idea of abjection to how the fecund female body is seen and felt in society.

With our emissions, undulations and folds, all of it magnified tenfold when we actually get down to the business of reproducing, we can seem a bit unbound. We are far removed from our vertical-silhouetted prepubescent selves – as yet unchanged by fertility – and even further removed from the clean lines of the male body. Our curves and fluids make us much more animal-like, which, if we follow Kristeva's thinking, could serve as a reminder of frailty and mortality. Death of the human body is the ultimate mess. To really consider it is to feel slightly scrambled inside.

Obviously, these are only ideas. They can help us understand our place in the world but we don't have to agree with them or feel able to apply them to our own lives. Because women's bodies are not abject. We are not constitutionally dangerous, unpredictable or spooky because we have bodies that grow babies and prepare for doing so every month. Our bodies have been *positioned* this way.

Hysteria: 'an animal within an animal'

'An animal within an animal' is what the Greek physician Aretaeus of Cappadocia considered the womb to be; an organ that 'moved of itself hither and thither in the flanks'.

In Ancient Greece it was believed that a woman's womb 'wandered' about the body. Women back then had the same working reproductive systems as we do now and, of course, will have experienced the same psychological and physiological changes in relation to their hormonal fluctuations throughout their cycles, pregnancies and menopause. Because hormones and female bodies in general were not well understood, women were, due to their mood swings and volatility, seen as irrational and incontinent. Agency-less against their 'bad' biology. Their erratic (a word that comes from the Latin verb *errare*, to 'wander' off course) behaviour was most often referred to as 'hysteria'.

In Victorian times, hysteria was closely linked to morality. When a woman was committed to an asylum in an extreme case of hysteria, her being taken away from public view with a diagnosis of a medical condition was viewed as an opportunity to maintain her nobility and reputation. Asylums were also seen as a way to restore the honour of 'fallen women'; a term used usually to describe women of the lower classes who had

had sex outside marriage and, often, prostitutes. Such women were believed to have lost control to sin and heathen desires.

Let's just remind ourselves, again, of how nebulous and subjective a diagnosis of hysteria was. The woman patient in a case of hysteria might exhibit any of a huge range of 'symptoms' including nervousness, insomnia, bloating, shortness of breath, trouble-making, sexual desire, faintness or irritability.

Most women will have their own associations with this word. Mine are profuse. Years and years and years of secretly carrying the gentle simmer of belief that I am just on the cusp of madness or losing control: of my emotions, my appetite, my sexuality, my fantasies, my body and its wet inner world. Sometimes, I think I'm a hair-trigger away from exploding. From releasing god knows what, of what nature, onto whom, and *why*, because I've quietly believed I'm . . . well, too much. Almost every woman I have known well has spoken of this belief; one they mostly keep tucked up, because admitting the fear means facing the idea that maybe, probably, we *don't* have inherent autonomy; that our woman-ness will always 'get' us in the end.

I'm quite tired.

The concept of 'hysteria', a word originating from the Greek word for womb, *hystera* (ὑστέρα), is described in the *Hippocratic Corpus*, the collection of sixty-odd Ancient Greek medical treatises associated with the teachings of Hippocrates, the man generally looked to as the founder of medicine as a rational science. Hippocrates was one of the first to identify this business of the 'wandering womb' and believed that hysteria, a catch-all term for most female illnesses and emotional excesses, anything from a headache to an epileptic fit to using a swear word, was a result of the womb actually *detaching* from the pelvic cavity and travelling around the body.

The natural extension of this belief was the idea that the womb must be confined, kept in place, in order to keep women stable and less dangerous. Furthermore, that the menstrual cycle was a fount of derangement; not just physical, but moral. Ancient Greek doctors prescribed all sorts to keep that pesky organ still. 'Pelvic massage' was the big one, involving the rhythmic spreading of warm, sweet-smelling oils over the whole external genital area, clitoris and vaginal opening, to lure the rambling womb back into its place. The practice likely produced quite a, er, therapeutic effect for women.

That women were orgasming all over the place while no one quite understood what they were made of inside is a source of comfort when you begin considering how we've been framed since, oh, the dawn of time. In her book *Women and Society in Greek and Roman Egypt*, the author Jane Rowlandson writes:

> The Hippocratics thought that the womb moved upward in the woman's body whenever it became hot and dry from overwork, or lack of irrigation from male seed, searching for cool and moist places in an effort to restore its equilibrium. As the womb tried to force its way toward the crowded places at the centre of a woman's trunk, it wreaked havoc with her physical and mental well being, causing her to faint or become speechless. Foul odors at the nose and sweet smells at the vagina were prescribed, to lure the uterus back to its seat.[33]

Hot and dry from overwork! Lack of irrigation from male seed!
It all sounds so bovine. Makes you think of knackered old cows being led around their paddocks by farmers desperate for them to just keep breeding. Perennially pregnant, leaking

and swollen. Sex and pregnancy were, for a long time, thought to be the only definitive cures for hysteria. It was the belief of medical professionals that, without a regular shafting, a woman's womb dries out and – alert the coastguard, close all the ports – is likely to become displaced. So you see we have quite a colourful history of our biology being misunderstood and, in turn, that biology being labelled as inherently 'bad'.

This misunderstanding had significant implications for women in society. Women were not part of Aristotle's teachings of philosophy, for example, because of his beliefs surrounding this 'bad biology'. He did not believe women could possibly be educated or become a part of the political landscape because, during the menstrual cycle, hormonal changes make them too prone to emotional changes. Unable to stay in control. In turn, how could they possibly have a sense of justice? Women were not allowed to sit on juries or be magistrates. They were outliers to the stable male norm, their physiology a marker of social status and, let's be clear, inferiority.

Hunting witches

Today, women lead. We have achieved glory and greatness in industry, politics, law, science, technology, space travel, art, music, sport and so much more in all corners of the world. We have *changed* the world. But has this association between our biological processes and our capabilities been completely rubbed away, particularly given that we're more aware of our hormonal changes than ever before? I'm not sure. To me it just looks different. In order to understand why, although it might not seem future-facing and progressive to do so, we must look at what's come before to understand where we still are now.

The naturalistic ideas of Ancient Greece gave way to an obsession with demonic possession in the Middle Ages, during which time it was thought that women prone to melancholy or any of the other symptoms of hysteria – fits, for example – were being controlled by demoniacal forces. If doctors couldn't identify the cause of a disease, it meant it was procured by the Devil. From the 1300s to the end of the 1600s, a witchcraft obsession swept through Europe. Tens of thousands of supposed witches were executed. Most of them were women. During the infamous Salem witch trials that occurred in colonial Massachusetts between 1692 and 1693, two hundred people were accused and tried for practising witchcraft. Sentenced to execution by hanging were five men, two dogs and fourteen women – all deemed to be displaying signs of possession. The difference in these numbers rings through the centuries.

The legal system we know today is a result of those with power in society looking at the witch trials and re-evaluating the way *all* trials are conducted. Eventually, it became evident that the right to a lawyer of choice, a sworn unbiased judge and jury, and working from the position of innocent until proved guilty were basic human liberties. However, the term 'witch hunt' is still used very often in reference to legal accusation and interrogation. It is a term mostly used by men when women come together to call out men's bad behaviour. Obviously, we think of Harvey Weinstein.

When Weinstein was arrested in spring 2018 for a catalogue of sexual assaults on women, including rape, it marked the moment that the #MeToo movement burst the dam. The levees – built from a lot of money and a lot of power – that

protected men like Weinstein from ever facing the consequences of their actions had been broken down by survivors'
testimonies. The foundations began wobbling over five
extraordinary days in October 2017, when two critical shots
were fired at Weinstein by two major American titles: first
by Jodi Kantor and Megan Twohey in the *New York Times*,
whose reporting detailed Weinstein's settling at least eight
times over nearly thirty years with women who had reported
sexual harassment and unwanted physical contact. Then,
in the *New Yorker*, Ronan Farrow laid out the details of the
alleged assault on actor Lucia Evans and twelve other women.
Kantor, Twohey and Farrow all, understandably, shared the
Pulitzer prize for public service for their work. Other women's
accounts began to mount and a terrible pattern started to
emerge. Again and again Weinstein had used his titanic
wealth, along with all the access to a legal arsenal that brings,
to silence his accusers. He is said to have promised future
work in exchange for silence, imposed gagging clauses in legal
settlements and planted calumnies about his accusers, heavy
on anonymous sources, in the tabloid press. He couldn't keep
what actor Salma Hayek called his 'machiavellian rage' contained[34], nor could he seem to control his cold, pathological
need to overpower women. A network of PAs, PRs, account-
ants and lawyers was employed to keep women's truths under
a stone. But they found a way out. They cracked the stone.

After those five days in October, more and more women
came forward about being assaulted by Weinstein. They
regained the power that had been taken from them. At the
time of writing Weinstein has pleaded not guilty to all counts
of non-consensual sex and his case is still ongoing. His career
is over. The motion-picture academy kicked him out. His
company is bankrupt. His wife divorced him. His political

allies have abandoned him. Whether or not his remaining time on earth will be spent staring at the grouting of a cell wall, the Weinstein case has changed things.

Weinstein might have been the eye of the cyclone but its perimeters continue to expand. The #MeToo movement has seen hundreds of powerful men in industries across the world brought down by allegations of sexual misconduct. We are witnessing a watershed moment in society; a seismic change in the power of survivor testimony in situations where power itself has been utterly abused.

Several commentators (most of them male) used the term 'witch-hunt' as the #MeToo movement gained momentum. That this archaic analogy is still used so often is quite fascinating to me. Speaking on *The Late Late Show* on RTE, actor Liam Neeson described the number of sexual misconduct allegations within the entertainment industry as 'a bit of a witch-hunt' and was roundly lambasted. The French actor Catherine Deneuve also encountered sharp criticism when she too used the term to describe what unfolded after the Weinstein case came to light. Woody Allen, he now of murky repute in his own right, commented that he was scared of a 'witch-hunt atmosphere, a Salem atmosphere'. (Allen later said Weinstein was a 'sick man'.)

When allegations of sexual misconduct against women – and men – arise, so too does this ugly phrase; more widely defined as 'mass hysteria'. Few men were persecuted in the Salem trials. The irony beneath the word choice is bitter. It speaks of an ugly history and has become a code for saying: women lie. It is used to dismiss serious allegations that challenge male positions of power. This is a systemic issue. No 'witches' need digging out. We must question the very *suggestion* that these common experiences are isolated enough to 'hunt' for in the first place. Men call serious sexual assault allegations 'moral panic', 'PC gone mad' or 'witch-hunts' because they are scared, on some deep level, of losing authority and power. As writer Lindy West said in an op-ed essay for the *New York Times* (deliciously titled 'Yes, This Is a Witch Hunt. I'm a Witch and I'm Hunting You'): when Allen and other men warn of a witch-hunt atmosphere, 'what they mean is an atmosphere in which they're expected to comport themselves with the care, consideration and fear of consequences that the rest of us call basic professionalism and respect for shared humanity'.[35] In the same piece, West reminds women that, although we don't have 'institutional power', we have our stories and we feel empowered to tell them. 'The witches are coming,' she said, 'but not for your life. We're coming for your legacy. The cost of being Harvey Weinstein is not getting to be Harvey Weinstein anymore.'

A huge malignancy is being reckoned with at last. To dismiss this with talk of hunting witches speaks of an urge to put a stop to accountability. Let's remember our history and create a different future. Let's remember how our justice system and wider society has a history of not believing women and call out this rot whenever we hear it. Language has real importance because what is actually at stake is women's safety.

Reputation

By using terms like 'witch-hunt' we are implying that women's safety has less value than the rusted reputations of Weinstein, Bill Cosby, James Toback, Roman Polanksi, Woody Allen and whoever else. We are *obsessed* with male reputation; the cult of artistic genius and what we should or shouldn't overlook of a person's conduct in light of the products of that genius. You know how the conversations go. In 2014, when the *New York Times* published an open letter from Dylan Farrow in which she first publicly shared details of the sexual assault she alleged her father, Woody Allen, made on her at the age of seven in an attic, many people in the public eye shook off the allegations as something private or unproved. Too murky to comment on. Farrow has been telling the same story for many years, each time eliciting vehement denial from Allen. Since the allegations have been part of public conversation, there has been a lot of, 'But: *Manhattan! Annie Hall! Hannah and Her Sisters!*' Since late 2017, however, in the beam of the #MeToo movement, the massive cultural reckoning around sexual abuse seems to have prompted much compassion for Farrow. For the first time this long untouchable legacy of Allen's has seemed to be under threat. Many actors have said they will no longer work with the prolific filmmaker. There is now an indelible toxicity attached to Allen, but it's taken time and he's still making movies.

As the writer Hadley Freeman pointed out in the *Guardian*, 'the Allen saga has always had a shape-shifting quality'.[36] When it became known, in 1992, that Allen was having an affair with Soon-Yi Previn, Mia Farrow's then 20-year-old adopted daughter, Farrow publicly accused Allen of assaulting their adopted daughter Dylan. 'The public response to this mess has altered with the times,' wrote Freeman. 'In the 90s people seemed weirded out more by Allen's relationship with Soon-Yi, to whom he has now been married for more than 20 years, than by the suggestion he might have molested a child.' Of course, people have been making grossed-out noises since *Manhattan*, Allen's 1979 film about his character's relationship with a 17-year-old schoolgirl, came out, but the case of sexual abuse allegedly happening within his own home is far from straightforward. Public opinion is not an impartial jury. Yet once you consider how much Allen's artistic *reputation* has been a source of protection, a blinker for those who love his films to this terrible thing having potentially happened, it does make you think more widely about what affects our judgement – and how easily – when it comes to women's accounts of trauma and pain.

We have a long history of assuming women's narratives are inherently untrustworthy. That history lingers in the fibres of society today. When a woman says she is in, or has known distress, it's like her words come with a pop-up ad saying: *Not to be taken at face value.* As the god Mercury warned so gravely in Virgil's *Aeneid*: *Varium et mutabile semper femina* ('woman is always fickle and changeable'). Feeling truly heard in the first instance when we talk about our distress, whatever form it takes, still feels like a rare thing for women; as though when we say we feel hurt, sad or scared, some kind of hysterical undertow is assumed. It cannot possibly just *be.*

The lack of power we can feel in discussing our experiences is the product of so many layers of patriarchal stranglehold it's hard to know where to start. So many systems are gathered in that fist. Political systems, healthcare, justice; in blunt and subtle ways, our experiences end up being refracted through male power of some kind. Sometimes we end up questioning or reframing what we know to be true for us, because life gets harder when we insist. It's often about reputation. We fear what will happen to our own and, in their positions of power, men fear what will happen to theirs.

I haven't spent much time really considering the forcefield of male reputation in my life. I have known what happens when men feel threatened, of course. I have seen the face of a colleague whom I reported for workplace bullying redden and shake as he furiously typed away next to me after we got back from a meeting with HR. I have seen incredulity in the raised eyebrows of a male GP as I tell him yes, for the fourth time, my periods really are *that* painful and I would like a referral to a gynaecologist, please. I have watched the faces of male friends and colleagues crinkle like newspaper when male artists they revere are accused of behaving badly. Sometimes, it takes a key voice to make you consider what binds all these experiences.

I have read feminist critical theory since I was an English Literature student at the start of the 2000s, but academia doesn't always stick. For me, nothing has skewered my experience as a woman quite like the Australian comedian Hannah Gadsby's stand-up special, *Nanette*, in June 2018. Gadsby's 45-minute set was the most powerful subversion of comedy I've ever seen; a steady deconstruction of joke-telling and how trauma can be cut off and frozen into pithy lines to get a laugh. But Gadsby also devoted much of her set to exploding the myth of 'separating the art from the artist' when men abuse their power, from Woody Allen to Louis C.K. to Pablo Picasso. Gadsby observes that these noted men are the rule, not the exception, and it is these men who nurture the popular belief that 'We don't give a fuck about women or children, we only care about a man's reputation.' She asks us to question their humanity. 'These men control our stories and yet they have a diminishing connection to their own humanity, and we don't mind so long as they get to hold on to their precious reputation.'

Towards the close of her set, Gadsby revisits a story she told early on about a man who threatened to beat her up at a bus stop for talking to his girlfriend: 'Oh sorry, I don't hit women, I got confused, I thought you were a fuckin' faggot trying to crack onto my girlfriend.' 'What a guy!' she laughs, along with the audience. But what actually happened has no punchline. The man came back. 'Oh no, I get it, you're a lady faggot, I'm allowed to beat the shit out of you.' She recounts the experience, no longer laughing. Her eyes shine with tears. 'And he did. He beat the shit out of me and nobody stopped him.' As she paints a picture of other traumas she has experienced over the course of her life, bound by misogyny and homophobia, Gadsby's audience is silenced. It's intentional. She wants them, and anyone watching at home, to marinate in the tension. Especially those who don't know what it means to go through life with a constant sense of apprehension, namely straight white men. 'This tension, it's yours,' she asserts. I had tears running down my face at this point. 'I am not helping you anymore.'

Although Gadsby may have seemingly dealt in the specifics of what came with being a gay woman in Australia's rural southern island state of Tasmania, *Nanette* felt strongly connected with the #MeToo movement and, as evidenced in women's responses to it across the world, hit the deep root of female experience. It spoke to any woman who has muffled or omitted the difficult bits of painful experiences for fear of making her listeners uncomfortable; for fear of disrupting the status quo; for fear of being seen a certain way, *that* way, you know what way: deviant, not normal, too emotional, too much to be palatable, not what a true woman *should* be; for fear of undoing years of female empowerment by 'giving into' the pain of past experiences; for fear of not being heard; for fear of the loneliness that rips you in half when you do speak and your words don't land but instead float towards the ceiling.

For fear.

Don't be so sensitive

Another stand-out bit of *Nanette* for me is when Gadsby addresses the 'Don't be so sensitive!' slur so often thrown at women when they're expressing emotion. I had a partner once who told me I really needed to grow a thicker skin to 'get on' in this world – a nugget of advice I tried to take on board for a while but, nah: I feel things.

'I get told to "stop being so sensitive" an awful lot. And it is always yelled. Which I find very insensitive,' Gadsby continues in her set. '"Stop being so sensitive." I don't understand. Why is insensitivity something to strive for? I happen to know that my sensitivity is my strength. I know that. It's my sensitivity that's helped me navigate a very difficult path in life. So when somebody tells me to "stop being so sensitive", you know what? I feel a little bit like a nose being lectured by a fart. Not the problem.' This spoke to me at spinal-cord level.

I've genuinely lost count of the number of times in my life I have witnessed noses being lectured by farts. Or, been the nose at the receiving end of the fart. I have been told I am too sensitive my whole life. But if being 'sensitive' means being open and responsive to other people's emotions, if it means being the friend that people come to when it really matters, when they're falling off the floor with despair or their child is sick, or if they have been catastrophically hurt by someone, then it's a privilege. It's taken thirty-four years and a lot of therapy to realise I can't really change my sensitivity and nor do I want to, because where the hell would I start? Which part would I begin to unthread first? It makes me who I am.

Since I was a small child I have delighted in the details of the natural world: veins in the sand at low tide, the tiny silver bobbles and yellow cups in patches of lichen, the hot velvet of a horse's nose, the handsome redness of a rosehip, the baby's-shoulder-like fuzz on a fig-leaf stem. In moments of real distress in my life, of which there have been a few, it has been the details of the bigger world around me, the riot of colour and life that exists and continues being what it is irrespective of what is happening in the world beneath my skull, that has brought the most comfort. I am a chronic noticer. Knowing that bushy odysseys of chlorophyll that'd make a watercolour-ist blush exist out there, rugged patterns in bark and sloshy mud and bodies of water and animals, birds, insects, all living their essence outside my front door, has always kept me going. Were I not so sensitive, I wonder if this would be the case. I don't care, really. I wouldn't change it.

I spoke to the novelist Charlotte Mendelson about this noticing thing. Mendelson's most recent book, *Rhapsody In Green*, her first non-fiction, is, on the surface, about an acci-dental gardening obsession and all the chaos that come with it. Beneath the delicious descriptions of leaves and stems, and the jokes about aphids, beans and slugs, however, is the spirit of someone who has, perhaps, always been noticing and delighting in the details of things as a way of connecting with a world that can feel tricky, alarming and embarrassing at times. Her award-winning novels (*Love In Idleness, Daughters of Jerusalem, When We Were Bad, Almost English*) have human sensitivity and awkwardness running through them like an artery. My hunch, it seems, was right.

'I've often been called sensitive, and it's never a compli-ment,' Mendelson tells me. 'It means, at least for a woman, too emotional, too touchy and quickly upset. That's probably true; certainly my life would be easier if I minded less. But if I wasn't so responsive, so alert to others' embarrassment, the little gestures which show their passions and fears, I wouldn't be a novelist.' And, if she was less aware of the world around her, Mendelson says she 'would be deprived of one of the joys of my life: the tiny details of nature, leaf-veins, stone speckles, bark and skin and tendrils. Being a noticer is a curse and a blessing, and it makes me me.'

How many women have been called 'over-sensitive' when displaying sadness, fear or anger? Told that they're over-reacting? As if the 'over' bit is not a wildly subjective conclusion depending on an individual's own thoughts and fears? All of us at some point, I'd imagine. A colleague of mine at VICE once looked at me with dilated cocaine pupils at the staff Christmas party and told me that, although he thought I was 'fucking amazing', I was 'too defensive' some-times. He said 'it really lets you down'. I went home soon after. I've heard it in pubs, living rooms, bedrooms, gyms, hospitals, you name it, and so, I suspect, will you have. It's usually from men. I've heard it come from the mouths of men I respect, including my own family, and men I don't respect. You hear women say it, too, and wonder what's going on for them, but it is usually men, isn't it? They think we either are too sensitive or have gone mad. But what if that 'madness' is, in fact, anger; anger with centuries of momen-tum behind it? I recall how my friend the writer Sophie Heawood closed a piece titled 'Princess Diana Was As Mad As Any Other Woman' about how bonkers people thought Princess Diana was ('The more dead that woman gets, the more I love her'):

> The older I get, the more I see how women are described
> as having gone mad, when what they've actually become
> is knowledgeable and powerful and fucking furious.[37]

97

After she tweeted the piece, the line was retweeted thousands and thousands of times. It's been in her bio for years now and, she tells me, 'may well be carved on my gravestone'.

Another tweet I saw that went viral – 18,000 retweets and 73,000 likes at the time of writing – was by Erin Keane, Executive Editor, Salon Media Group:

> Every woman I know has been storing anger for years in her body and it's starting to feel like bees are going to pour out of all of our mouths at the same time.

These women are right. Of course they are. Other women have shared their sentiments in droves because we *know* it. Our cells are full of it. Our opinions, emotions and behaviour have been pathologised and written off as unnatural, women positioned as excessive, volatile and downright untrustworthy, whether or not the person saying it actually believes that in their bones, for so long. It's so convenient, so fucking *easy* to play the woman-is-monster card when you don't want to listen to what she's saying. Or, don't have the capacity to. The tendrils of stigma, shame and otherising from the 'hysteria' days still poke and prod today. It's a vapour we all breathe, day in, day out. Let's just call it what it is: something not very decent. Something that says more, so much more, about the person saying it. At a push we could say it's gaslighting. 'You're too sensitive' is what people say when they've said or done something unkind and want you to believe that they haven't. It's gaslighting because when you hear it often enough, the possibility that it might be true spreads over your mind like a bad fungus.

Exploding

In recent years it has felt like we've been having a reckoning; one that's been hard to define but which, of course, raged like a tsunami with the #MeToo movement. Another part of it, for me, has been about seeing realistic portrayals of women on our TV and cinema screens – inevitably those written by women. Phoebe Waller-Bridge's *Fleabag* comes to mind. *Fleabag*, which began life as a fringe play at Edinburgh, is a dark, filthy comedy of many layers. It is, at first glance, the story of a city-dwelling young woman in that painfully familiar territory of feeling like she's taking too long to figure out who she really is and what she really wants, being a bit of a dick along the way – to others and herself. Very quickly, though, Waller-Bridge's six-episode series unfolds into a ferocious dissection of grief, memory, trauma, friendship, family dynamics, self-esteem and the pain of love. The seam that binds this emotional patchwork is female anger. This shouldn't feel revolutionary but it does.

In the first episode of *Fleabag* there's a line that perfectly skewers the mood of the rest of the series. Fleabag – we don't know her real name – is in the art studio of her pass-agg stepmother (Olivia Coleman) late at night and says, in a confiding aside to the camera, 'She's not an evil stepmother, she's just a cunt.' The consonants of the c-word are like bullets. She smiles tightly. She's livid.

In an interview around the time of the show's release on BBC Three, Waller-Smith said, '... I know a lot of my female peers feel really angry. I think that a woman's response at first is to feel guilty and apologetic about it without knowing why ... The idea of the "angry young man" is so deeply embedded [in culture] but the angry young woman seems never to be addressed.'[38] With *Fleabag*, she did it. I wrote a piece about the show for the *Guardian* that, for a few days, was the most-read piece on the whole website. There is appetite for a rage revolution.

For years, it feels like mainstream TV has shied away from female anger unless it was victim's rage or the simple – dreaded – female 'hysteria': the crazy gal who can't hold it together. Women of Fleabag's age (we're never told how old she is, but Waller-Bridge is thirty-three) grew up with shows like *Friends* and *Sex and the City*, too, in which the anatomy of female friendships was explored at length, but in which a woman's individual anger was scripted to be saved up and shared with a best friend. The safe spaces for emotional jettisoning in *SATC*, the real, raw stuff, were the brunch table or a funky Midtown bar – not pavements, like we see in *Fleabag*, pounded alone with hot, angry tears dripping black lines down her face. The pockets of despair, both quiet and loud, come with no apologies.

Fleabag didn't deserve to be pitted against any other shows written by women of a certain age – *Girls* comparisons were plenty – but the woman Waller-Bridge brought to life on our screens is, it seems, part of a steady sea-change of television writing that allows the complexities of female characters to stretch and breathe.

Although they are very different shows (one packs far more one-liners about anuses), I got a similar feeling watching *Fleabag* as I did watching the first season of Jane Campion's magnificent *Top of the Lake*. Robin Griffin (Elisabeth Moss), the Sydney police officer returning to her remote hometown of Laketop, New Zealand, to investigate the pregnancy and then disappearance of a twelve-year-old girl, Tui, from her rural home, almost vibrates with tension. In one episode, she sinks a dart into a man's back in the middle of a dive bar, angry at the way he is talking about the Tui case — an act made more violent by her casualness. But then, as we discover, Robin has a lot to be angry about. She is enraged by sexual violence and the prospect of what could have been done to Tui, the endemic misogyny within both her home-town community and the local police force, her strained relationship with her fiancé back home, her mother's cancer and the gradual hinting of a childhood trauma involving some of the local men she's having to deal with now. *Top of the Lake* was astonishing for many reasons, but the ways in which Robin's anger, tall as the mountains of the South Island surrounding her, was made so central to the storyline was defiant. The sharp edges of female emotion glinting on screen give such a thrill.

Through the strength of the script and the actors playing them, these female characters regularly unleash rage, and it's not always righteous. Sometimes it's un-pretty and without clear reason at the time. But this is what it's like for so many women, who live and breathe a lifetime of cumulative anger but feel as though they should keep a cork in it for fear of the seepage being seen as a character defect. That is why these women are so thrilling to watch. Waller-Bridge did it again in 2018 with the Emmy-award-winning *Killing Eve*, her glorious thriller series based on Luke Jennings's *Codename Villanelle* novella series. Ridley Scott described it as a threat to the movie industry. With it, Waller-Bridge shoves a 10,000-volt shock up the rear of the well-worn game of cat and mouse staged by so many male-heavy murder mystery series.

In *Killing Eve*, both the spy (Eve Polastri, played by the untiring and very funny Sandra Oh) and the dexterous assassin (Villanelle, played with fierce wit by Jodie Comer) are women. One of the funniest things about the show is that you often find yourself wanting to be, kiss or at least admire Villanelle, in spite of all the throat-slitting, shooting and watching-people's-last-breaths-with-a-child-like-smile-on-your-face stuff. In an interview with the *Guardian*, actor Fiona Shaw, who plays Carolyn, the MI5 boss overseeing the quest to find Villanelle, says that the way *Killing Eve* flirts with amorality is why it's so powerful: 'There's no virtue in it, so it's women not being virtuous,' she says. 'It's fantastic to have an antiheroine like Villanelle, doing all the things you might think, but never dare to do. And she succeeds! That's really exciting. It's like liking the devil, isn't it? But there's no consequence to it. She always comes out fine, and everyone dies. I think it's playful, and really anarchic.'[39]

Women have been watched by people waiting for them to explode for so long. Now, it tentatively feels like we're starting to explode on our terms.

Lord, but have we been watched, though.

Surveillance

As Michel Foucault argued in Volume I of his infamous study of sexuality in the Western world, *The History of Sexuality*, since the beginning of the eighteenth century medical experts have subjected the female body to scrutiny and surveillance at the physical and biochemical level. This has positioned the female body as an inevitable place of badness, madness or weakness. As such, a tyranny of 'truths' as to what is normal or abnormal in terms of female sexuality has persisted over time.

The eighteenth century, also known as the 'age of enlightenment', was a pivotal time for medicine. The business of human bodies and how they are maintained, as well as how we understand and treat them when they go wrong, began to become mainstream. Scientific knowledge slowly punctured public discourse. The origins of many diseases and disorders were discovered, along with potential antidotes. The religious edicts of the seventeenth century, along with the rituals associated with the fertile female body and stemming its excess, began, slowly, to be shaken off. Yet not entirely. We only have to look at how religious rituals around menstruation still exist in many cultures across the world today to see how people are still trying to quell this horror associated with abjection and regain 'purity'.

According to World Health Organisation statistics, in our twenty-first-century world 200 million women and girls don't have access to adequate sanitary care. There is still so much stigma and misinformation surrounding menstruation in Iran that 48 per cent of girls there think that it's a disease. In Bolivia, girls are urged, even by teachers, to keep their used sanitary pads far away from the rest of the household rubbish because traditional beliefs hold that disposing of their pads with other garbage could lead to sickness or cancer. A UNESCO report in 2014 estimated that one in ten girls in sub-Saharan Africa misses school during her menstrual cycle.[40] By some estimates, this equates to as much as 20 per cent of the school year.

Consider, too, the Jewish law of niddah. In the book of Leviticus, the Torah prohibits any sexual intercourse with a niddah – the word used to describe a woman having her period, or a woman who has had her period and not yet, as required, immersed herself in a mikveh (ritual bath). Other regulations include sleeping in separate beds while a woman is niddah, and, for some Orthodox couples, avoiding seeing one another naked.

I haven't once read or thought about opening a Bible since Religious Education lessons at school, but, for research purposes, I was compelled. It struck me as important to try to understand the genealogy of this fixation on uncleanness. It is alien and archaic to me, a white atheist living in the Western world, but it is not so to millions of people.

And if a woman has an issue, and her issue in her flesh be blood, she shall be seven days in her menstrual separation: and whoever touches her shall be unclean until evening. And everything that she lies upon in her separation shall be unclean: everything also that she sits upon in her separation shall be unclean. And whoever touches her bed shall wash his clothes, and bathe himself in water, and be unclean until the evening. And if it be on her bed, or on anything whereon she sits, when he touches it, he shall be unclean until the evening. And if any man lies with her at all, and her menstrual flow be upon him, he shall be unclean seven days, and all the bed on which he lies shall be unclean ... But if she be cleansed of her issue, then she shall number herself to seven days, and after that she shall be clean. (Leviticus 15:19–29)

The reign of the Church began to loosen in the eighteenth century. However, any medical text from the time will tell you that female bodies were – despite their constant dissection and scrutiny as the bearers of babies and therefore, you know, important to understand so the human race could be kept going – largely seen as inferior male bodies. That 'bad biology' theory of Aristotle's still hung heavy in the air. Progressive scientific views may have held men and women to be largely equivalent in terms of their bones, muscles, brain structure, organs and nervous systems, but women were *still* determined by their biological role – mother – and the organs that allowed them to fulfil that role. We should keep it in mind, too, that medicine was an exclusively male profession; men who had privileged, unrestrained access to women's bodies. Women were cut up in operating theatres and locked up in institutions because men thought they were too much: too emotional, too sexual, too messy, too loud, too sad, too frustrated, too demanding of them and wider society. Male doctors didn't just want to understand *why* women seemed sad or mad, either; they wanted to immediately lessen their impact on those around them.

Living as a woman in the nineteenth century was no walk in the park. Victorian society emphasised female purity and the ideal of the 'true woman' as wife, mother and house-keeper. As guardians of home and family, women were believed to be more emotional and gentle of nature. This perception of femininity fed the popular idea that women were inherently more vulnerable to illness and disease – particularly hysteria; the basis for a diagnosis of insanity in so many female patients. Men were diagnosed with hysteria too, but not nearly as often, it would seem. Assigning the condition to female nature fitted the Victorian model of women who, in the middle and upper classes at least, were utterly dependent on their fathers and husbands. With little room for power, control and independence in the day-to-day particulars of being a respectable home-keeper, daughter and mother under the muscle of the patriarchy and society's gender ideologies, perhaps it's no wonder that women were struggling to cope.

Asylums were public institutions. As such, they were watched and critiqued by the general public. Opinion was

105

very important. Those running the institutions knew the importance of being seen as innovative and progressive in terms of treatments. The London Asylum for the Insane was opened in Canada in 1870 and was run for a time by a superintendent named Dr R. Maurice Bucke. He believed that women's reproductive organs were inextricably connected to mental illness. As such, treatments Dr Bucke adopted at the London Asylum included hysterectomy, because he believed that removing a woman's harrowed womb would restore her sanity. In an article titled 'The Evolution of a Mystic' from a 1966 edition of the *Canadian Journal of Psychiatry*, it's stated that, despite criticism from other doctors that his procedures were 'meddlesome' and 'the mutilation of helpless lunatics', Bucke continued practising hysterectomy as a treatment for hysteria at the London Asylum until he died, in 1902.[41]

Our colourful history of diagnosing women as hysterical and putting them in psychiatric hospitals is something psychiatrists remain mindful of today because how could they not? Male dominance was the cornerstone of Victorian psychiatry. The proliferation of asylums between 1800 and 1900 and the number of women inside them meant that very few female voices were informing politics and, therefore, what autonomy women had. A few stuck their heads above the parapet, though. An American woman called Elizabeth Packard was a notorious social reformer who, after three years in an asylum – committed by her husband in 1860 because he deemed her insane for questioning his religious and moral beliefs: something he could do then without public hearing or consent – began campaigning for protective legislation for the insane and improved rights for married women. Another important figure in the reform of mental healthcare was

Dorothea Dix, one of the most influential asylum reformers of the 1800s. Dix played an instrumental role in founding or expanding more than thirty hospitals for the treatment of the mentally unwell and spearheaded the movement challenging the idea that those who were mentally distressed could not recover or be helped.

These brilliant women helped change the tide, yet the ocean remained murky. It took until the late 1800s for 'hysteria' to become the focus of intense medical attention, rather than being thought of as an exclusively female affliction. In the meantime, it was not just wombs that male doctors were cutting from women's bodies.

Bad clitoris!

One morning while researching Victorian medical literature in the British Library I discovered that male surgeons were also removing women's clitorises. I half expected to see my own grey matter splatted across the nice jumper of the person sitting in front of me.

Clitoridectomy, the surgical removal of the clitoris, was briefly thought an acceptable treatment for hysteria. A man called Isaac Baker Brown, a prominent nineteenth-century English gynaecologist and obstetrical surgeon, became a fellow of the College of Surgeons in 1848 and developed new operations to treat ovarian cysts and tumours. In 1858 Baker Brown set up a clinic in Notting Hill called – adopt brace position – The London Surgical Home for the Reception of Gentlewomen and Females of Respectability suffering from Curable Surgical Diseases. Here, he performed what he would simply call 'the operation'. Baker Brown had all kinds of theories about women's mind–body connections. In the second edition of his book *On Surgical Diseases of Women* (1861), a condition called 'Hypertrophy and Irritation of the Clitoris' was included.[42] Baker Brown believed this 'irritation', also referred to as 'the peripheral excitement of the pudic nerve', could have significant effects on a woman's nervous system and make her infertile. Essentially, masturbation equalled madness. A woman's inability to control herself equalled spinal problems, deformities and, ultimately, death. These women were coming themselves into early graves. But, lo: there was a cure!

Standard treatments at the time for women's 'irritated' genitals and associated 'mania' included filling the vagina with water and ice or applying leeches to the labia so they could literally bite and suck the life from our most vascular, tender parts. The wounds would seep for hours. Days. Baker Brown, however, thought the answer lay in something more definitive. Carrying out this surgery between 1859 and 1866, he operated on women with epilepsy and, chillingly, on girls as young as ten. It is almost impossible to imagine how the conversation between a ten-year-old's parent and a doctor proposing such a thing goes. He operated on five women whose 'madness' was them wanting to divorce their husbands and reported that, in each case, the woman returned 'humbly' to her husband after her clitoris had been removed. Baker Brown eventually began to receive negative feedback from others in the medical profession, questioning his success claims and arguing that he was operating both on women of unsound mind and those who had not given their consent. He was expelled from the Obstetrical Society of London in 1867; barely 150 years ago. We now so condemn female genital mutilation (FGM) as a barbaric, inhumane act of cultures 'other' to our own, that to find it happening in Victorian Britain is quite astounding.

The clitoris was an important part of physicians' understanding and 'treatment' of hysteria throughout this period. When doctors were not cutting it off a woman's body, they were actually manually stimulating it and making women come in order to rid them of their wretched affliction. Not that pleasure was an implicit part of the deal. Women in the Victorian era were not supposed to feel sexual desire. Or, if they did, it should only be when a man's penis was inside them. Hysteria was considered a disease utterly removed from sex. The predominant cure – if the woman was not committed to an asylum – was, as had been the case since the Ancient Greeks, pelvic massage.

Clitoral stimulation was thought of as the palliative treatment for an illness. Women took themselves to their doctors, or, more likely, were marched there by their husbands, to be cured of whatever malady fell within the diagnostic brackets of hysteria. So, in their offices, doctors were doing these 'pelvic massages' and bringing women to orgasm all over the shop. Yet it wasn't *called* orgasm: if a woman's face became flushed and she seemed relieved, happy and lighter after her massage, she was said to have experienced a 'hysterical paroxysm'. Snappy little phrase.

In her book *The Technology of Orgasm: 'Hysteria,' the Vibrator, and Women's Satisfaction* the American scholar Rachel Maines hypothesises that doctors practised with 'the comforting belief that only penetration was sexually stimulating to women. Thus the speculum and the tampon were originally more controversial in medical circles than was the vibrator.'[43] If a woman expressed desire for her clitoris to be stimulated with a vibrator, then, she was clearly diseased. Naturally, these so-called hysterics would only be 'cured' for a while. They'd return again and again seeking release – the kind, you imagine, they were not getting at home.

When the Industrial Revolution came, so to speak, and household appliances became electrified, vibrators entered the home and women could see to themselves privately. Devices were advertised widely in women's magazines as 'massage wands' yet no one spoke about their sexual function. That changed in the 1920s. Vibrators began appearing in stag films (the silent, secretly produced pornography of the era) and their function of giving sexual pleasure became increasingly hard to ignore. Magazines gradually stopped selling advertising space to vibrator manufacturers.

Interestingly, the first battery-powered vibrator was invented in the 1880s by a man called Mortimer Granville, who was actually vehemently against it being used on the clitoris – he called it a 'percusser' and intended for it to be used to relieve muscular aches and pains. Granville desperately tried to distance himself from the 'mis-use' of the device by physicians trying to make women get a move on with their paroxysms. In his 1883 book *Nerve-Vibration and Excitation as Agents in the Treatment of Functional Disorder and Organic Disease*, he wrote:

> I have never yet percussed a female patient ... I have avoided, and shall continue to avoid, the treatment of women by percussion, simply because I do not want to be hoodwinked, and help to mislead others, by the vagaries of the hysterical state or the characteristic phenomena of mimetic disease.[44]

In other words, our friend Mortimer thought women were malingering in order to get off.

Given how commonplace pelvic massage became as a 'treatment', women clearly knew what needed to be done to which bits of their body in order to have an orgasm. Yet women's sexual pleasure, particularly that which is experienced via the clitoris, was both an obsession and source of great vexation for male doctors. We know that Sigmund Freud, a neurologist, was fascinated with hysteria. The most notorious case study in the history of psychoanalysis is the case of Anna O, the first in Josef Breuer and Freud's *Studies on Hysteria* (1895), one of the most famous publications in psychology. The case concerns a patient, real name Bertha Pappenheim, who sought Breuer's help for headaches, visual disturbances, hallucinations, partial paralysis and speech problems. He diagnosed her with hysteria and began visiting her every day. Breuer noticed that her condition improved each time he gave her the opportunity to talk at length, emptying her mind to him and releasing her anxious thoughts. They continued this dialogue from December 1880 to June 1882. Pappenheim herself coined the phrase 'talking cure' and her case is often referred to as the founding of psychoanalysis. Her treatment led to greater understanding of the impact of past traumas and subconscious memories on the conscious mind, along with practices like regression and hypnosis as means of identifying possible causes of mental distress. Freud referred to it again and again throughout his career.

One thing Freud seemed opposed to accepting in his studies of hysteria was female pleasure through the clitoris. Yeesh, did our relationship with these little organs of ours – originating from the same developmental tissue as the penis and designed purely *for* pleasure – weigh heavy on his mind. In 1905, Freud declared that women who orgasmed via clitoral stimulation were immature: 'Elimination of clitoral sexuality is a necessary precondition for the development of femininity, since it is immature and masculine in its nature.'[45] He believed that, after puberty, women should only be having penetrative vaginal orgasms. The theory had no basis in observation or, indeed, anatomical understanding. Yet this notion – that there is a proper, 'mature' orgasm squirrelled away inside every woman's vagina and that it is up to her to find it – *still*, despite everything we know about female pleasure, lingers in modern society. Imagine men being told they should stop savouring their stupid old penises! To stop messing about and, somehow, start having orgasms *only* via their anuses! And it's the asylum for you, dear sir, if your bum doesn't work!

'Returning' to the sexual body

When Naomi Wolf's book *The Beauty Myth: How Images of Beauty are Used Against Women* came out in 1991, many felt she had started another wave of feminism in the ways she made modern the old idea: that women were still being reduced to their bodies. In 2012, Wolf brought out a new book, *Vagina: A New Biography*, sold as a treatise on how significantly a woman's bodily experience influences most aspects of life. Its writing came about after Wolf had had an epiphany born of a 'medical crisis'. Where having an orgasm had once led her to see 'colours as if they were brighter', she felt she no longer experienced sex in a 'poetic dimension ... instead, things seemed discrete and unrelated to me'. Wolf went to her gynaecologist and then to a doctor (her 'nerve man') who told her she had a mild form of spina bifida. Her spinal column was a little wonky and, he said, was compressing the arm of her pelvic nerve that ends in the vaginal canal. The nerve man then operated on her, fusing her vertebrae as a means of releasing pressure on this nerve. It was a success. 'I began again, after lovemaking, to experience the sense of heightened interconnectedness,' she writes 'which Romantic poets and painters called "the Sublime".' Wolf's nerve man had spent a lot of time telling her about how pelvic nerve impulses travel up to the 'female brain' and that, because every woman's physiology is different, our ability to have rip-roaring orgasms mostly depends on neural wiring. Not the culture we live in, historical conditioning or anything like that. Hm.

Critics tended to fall in one of two directions when *Vagina* came out. In a piece on her website Brain Pickings, the writer and critic Maria Popova described *Vagina* as 'a fascinating exploration of the science behind the vastly misunderstood mind–body connection between brain and genitalia, consciousness and sexuality, the poetic and the scientific'. In the *New Yorker*, the writer Ariel Levy wasn't so taken. In her review, 'The Space in Between', Levy writes that many feminists 'may be perplexed to find Wolf, in her eighth book, situating the essence of the female being right back where it started: in the body, in one particular place'.[46]

Certainly, when I read the book, I found Wolf's veneration of an investment-banker-turned-sexual-healer called Mike Lousada a little … troubling. This Lousada specialises in 'yoni massage' and 'yoni-tapping' – treatments that, as Wolf describes it, 'address the trauma stored in the genitals'.[47] He often does this naked, but stipulates that he doesn't 'generally have intercourse with my clients unless it is extremely therapeutic'. (Doesn't *generally* have intercourse with his clients. How phlegmatic of him.) My favourite detail was how Lousada claims he once 'actually had an experience of seeing the Divine within the vagina', saying that 'an image came to me of the Virgin Mary'. As Levy rightly identifies, not all vaginas want the same thing: 'Wolf forgets that vaginas – like their handmaidens, women – have distinct personalities and preferences. If my vagina heard a potential partner murmur, "Welcome, Goddess," she would turn to me and say, "Get us out of here now."' I think mine might set itself on fire.

So Wolf may have invited women to try a new thing in directing the site of our consciousness back between our legs. Or at least give the thing a bit more attention as a kind of air traffic control centre for the rest of our being. The conversation that happened around *Vagina* was interesting because at its centre was the question of whether returning to the sexual body does us any good or not. The dusty relics of history can be shocking, maddening and, sometimes, hilarious. Do we run the risk of being reductive by talking about history's ignorance and many control tactics – wilful or otherwise – surrounding the female sexual body; the locus of radical feminism's last fight? Should we leave it alone? Depends on where you set your parameters of The Past, I suppose, because doctors were still saying up until the 1960s that women who needed clitoral stimulation in order to have an orgasm were mentally disturbed. A lot can happen in fifty years, sure, but it's not *that* long ago. In his book *The Sexually Adequate Female*, published in 1964, the psychiatrist Dr Frank S. Caprio writes: 'Whenever a woman is incapable of achieving an orgasm via coitus, provided the husband is an adequate partner, and prefers clitoral stimulation to any other form of sexual activity, she can be regarded as suffering from frigidity and requires psychiatric assistance.'[48]

Today, we know a lot more about the anatomy of the clitoris. We'd be wrong to assume that all there is to it is the super-sensitive little pyramid of flesh between the labia. In 2005, an Australian urologist named Helen O'Connell published a comprehensive anatomical study titled *Anatomy of the clitoris*[49] and, with it, began to change our understanding

114

of female sexuality. In her huge review of studies spanning dissection, micro-dissection, magnetic resonance imaging (MRI), three-dimensional sectional anatomy reconstruction and histology, O'Connell established that the whole clitoris is, in fact, a wishbone-shaped organ about 3.5 inches long and 2.5 inches wide. It looks a bit like a seagull in mid-flight, its wings in a downwards flap. The 'glans' of the clitoris is just the visible tip; the organ extends into the body and splits into two leg-like pieces, the crura, adjacent to the vagina and urethra. O'Connell uses the phrase 'neural trunks', which I just find so pleasing and can't stop saying. 'The vaginal wall is, in fact, the clitoris,' she writes. 'If you lift the skin off the vagina on the side walls, you get the bulbs of the clitoris – triangular, crescental masses of erectile tissue.' That mysterious, long obsessed-over 'G-spot' at the front of a woman's vagina is, in fact, part of the clitoral structure.

Plenty of research is still to be done in understanding how the clitoris connects with other structures in the body, as well as the nervous and circulatory systems, but we can flick a confident two fingers in Freud's direction because we now know that parts of the clitoris are hidden both in the vagina and near the urethra and anus. So, even if a different part of the clitoris is stimulated during penetration than in masturbation, the pleasure is still clitoral.

Come again?

An orgasm might feel specifically located in our bodies, but those brief moments of ecstasy we (hopefully) experience during sex or masturbation are actually a combination of neural, cognitive, emotional, somatic (relating to the body) and visceral processes.

Genital stimulation sends strong signals to the limbic system, the part of the brain that deals with emotion, memories and arousal. This system includes the hippocampus, which is where our memories and fantasies 'live'. That it is activated when we're aroused serves as some explanation as to why images of exes, colleagues, characters in books, that person who served you a coffee this morning, etc., etc., might keep popping into your head while you're on the path to orgasm. I have a friend who told me that she always thinks about a particular dry-stone wall in the Cotswolds as she's about to come: a funny yet fascinating insight into brain topography. She thinks that, at some point in her early orgasming life, an image of that wall popped into her head while she was ascending. Now her brain associates the image with the pre-orgasmic phase, her hippocampus holding it up like a cabbie with a name sign at an airport. Sex can be an incredibly emotional experience. Sometimes we cry. Given that the limbic system also includes the amygdala, which has a central role in our emotional processing, the high-emotional state makes sense. There is a sense of primal need and purpose in the surrendering of our own body to another, but also vulnerability. In that surrender we need to process an element of fear, even if it's not something we can consciously attach words to.

Interestingly, brain imaging studies have shown that orgasm causes a marked decrease in activation of other brain regions associated with fear, behavioural control and anxiety. The frontal cortex is highly involved in moral reasoning and social judgement, both of which are suppressed during orgasm, leading to absence of moral judgement and self-referential thought. Perhaps this is why we so often talk about 'losing' ourselves during sex. Increased activity in two areas called the anterior cingulate cortex and the insular cortex that are associated with the brain's pleasure response might help inhibit the sensation of pain for some people, or make a certain level of pain – hair-pulling, biting, scratching, squeezing, smacking; whatever melts your butter – pleasurable during sex. There is evidence to suggest that women's pain thresholds more than double during orgasm and the build-up to it.

The hormone and neurotransmitter oxytocin is produced by the hypothalamus during orgasm. Levels increase rapidly during sexual arousal and peak in a big burst at orgasm. In fact, some research shows that the intensity of an orgasm correlates with the amount of oxytocin produced. In other words, the more oxytocin, the harder your legs will tremble. Oxytocin levels have been found to fluctuate during a woman's menstrual cycle, being lowest in the luteal phase and highest in the ovulatory and follicular phases. We often hear oxytocin referred to as the 'love hormone' or the 'cuddle chemical' as it's known for producing feelings of wellbeing, reduced anxiety, and increased trust.

Our post-orgasmic chemical marinade also includes a flood of dopamine, shown by a large body of evidence to be the key neurotransmitter involved in making orgasm happen. The release of dopamine is activated by the nucleus accumbens — the reward centre of the brain — to 'reward' us for the sex. This is the same area activated by narcotic drugs, alcohol, nicotine, caffeine and chocolate; all the addictive things, potentially explaining why sex and the need for sex can feel like a drug. Recent literature also suggests that self-harm and suicidal behaviour can be conceptualised as addictions because of how the dopaminergic 'reward' system is activated in the process.

There are lots of brain-imaging studies now that show

how the dopaminergic 'reward' system is activated during sexual arousal and orgasm. They are mostly fMRI (functional magnetic resonance imaging) scans, which are based on the same technology as normal MRIs – a non-invasive test that uses radio waves and a very strong magnetic field to create detailed images of the body – but which look at blood flow in the brain rather than at organs and tissues. Increased blood flow is associated with 'neuronal activity', which basically means nerve cells getting excited and starting to do stuff. All fMRI scans are gorgeous to look at, because of how areas of increased blood flow glow from within black-and-grey images of the brain, like Christmas lights in the dark, but scans of brains during orgasm, easily google-able, are spectacular: big, blowzy waves of reds, yellows and oranges, like blooms of amaryllis, dianthus and begonia. That these images were dependent on a participant being able to either masturbate or be stimulated by their partner to the point of orgasm inside a scanning machine, with researchers on the other side of a thin pane of glass, makes them even more remarkable. Hopefully decent modesty curtains were erected.

Oxytocin

In women, oxytocin is most well known for its role in birth and lactation. Most notably, it is released at the end of foetal development when the baby is ready to be born, prompting the uterine contractions that will release the baby from your body and into the world. Sometimes when I stop and contemplate this pulling apart of two bodies, one tiny, naked and prawn-pink, one big, stretched and exhausted, still joined together by this silvery cable of vitality, it makes me light-headed.

The way oxytocin works during childbirth is fascinating. It is released during a positive-feedback loop. The exiting baby is huge and heavy, pressing against the cervix. The muscles of the cervix respond to the increasing pressure by stretching open bit by bit, sending nerve impulses to the brain. The brain releases oxytocin, which continues to dilate and soften the cervix. With more room, the baby's weight pushes in a downward motion onto the softened cervix, stretching it more. The function of oxytocin triggers the release of *more* oxytocin. Oxytocin is so integral to childbirth that a synthetic drug mimicking its chemical structure, Pitocin, is used when labour needs to be induced. Medication that blocks the action of oxytocin (like the drug Tractocile) is used to halt premature labour because without oxytocin's stimulus the uterine contractions cease.

Oxytocin's other function is in lactation, serving a very important role in milk letdown. This is the case for most female mammals, not just humans. Other hormones like prolactin are involved in the actual *making* of milk, but oxytocin is fundamental for releasing it to nourish the baby. The simplified process goes like this: baby sucks at mother's nipple; the suckling sensation sends signals to the brain; the hypothalamus is stimulated and oxytocin is released; oxytocin helps contract the muscles around the alveoli (the grape-like clusters of tissues that secrete milk) in the breast; the squeezing of the alveoli pushes milk into the milk sinuses; baby suckles more and gets the milk. I am reminded, as I write this, of the first time I went to visit my best friend Kate after she'd given birth to her now primary-school-aged daughter on her living-room floor. As I walked through the door she squeezed her nipple at my face, sending an arc of cotton-white liquid squirting across the bridge of my nose. Later, after we'd laid out this tiny, pouty and snuffling grub of a new person on the bed for a full inspection of her arm creases, soft shoulder pelt and ten juicy toes, Kate took off her pyjamas and asked me what her naked body looked like after the event. All I could think about was what it had done just a few days ago and, in that moment, I thought that her form was probably the most powerful thing I'd ever seen.

Nipple sadness

This milky research made me think about something I've only really discussed with my closest friends, some of whom identify, some of whom have said, '*What?*': sometimes, something coming into contact with my nipples – for example during sex, brushing too hard against a top, getting caught in a sports bra, drying after a shower – can make me feel . . . sad. But only for a few moments. It's so strange, this feeling, and doesn't happen all the time. I remember discussing it years ago with another of my closest friends, Nell, who said she had exactly the same thing, describing it as 'like landing on your crossbar', which is exactly it: deep, abstract and weirdly embarrassing. At the time of writing Nell is nursing her beautiful ten-month-old baby boy. She tells me that she felt 'that crossbar feeling' significantly at the beginning of breastfeeding and that it was something she had been warned about by other mothers. When her partner first asked how breastfeeding felt, she tells me she just replied with 'really sad'. 'On the day my milk came in, whhoooooeeeeeee. [I quote verbatim from our WhatsApp chat box.] It only lasted a few seconds but was so intense. I sobbed and sobbed.' I'd had no idea. My poor, lovely friend. Talking more about this with Nell, I discovered that this feeling has a name: dysphoric milk-ejection reflex, or D-MER for short. I had never heard of it before.

D-MER only starts appearing in medical research as a recognised condition affecting lactating women at the beginning of the 2000s. It is characterised by an abrupt sense of dysphoria: the etymological opposite of euphoria, reported by women in terms ranging from homesickness and sadness to dread, anxiety, paranoia, irritability and restlessness. It can occur just before the milk is released and usually only lasts for a few minutes. I have read through online forums of women who talk about feeling it so intensely that they switch to formula or begin weaning early. D-MER is a physiological response to changing chemicals in the brain, rather than a multi-causal problem like depression or generalised anxiety, so having a history of mental health problems doesn't appear to be a risk factor. However, if you are feeling moments of intense dread while breastfeeding your baby, it is bound to provoke more general feelings of anxiety or trepidation about the act itself. As a term, D-MER is often found through Google searches like 'sadness when breastfeeding' – imagine all those anxious mothers on sofas across the world, holding their nursing babies in one arm, typing into Google with their free hand – and information is spreading through increased online content and community awareness.

There is scant research on D-MER at the moment but preliminary anecdotal evidence shows that D-MER is treatable if severe, and initial investigations suggest that an inappropriate drop in dopamine at the time of the milk ejection reflex is the cause. Dopamine inhibits prolactin, the hormone involved in making the milk. (Prolactin also has other wide-ranging functions in the body for men and women, from acting on the reproductive system to influencing behaviour and regulating the immune system. Too high prolactin levels in men can cause testosterone deficiency syndrome, characterised by mood changes, the growth of breast tissue, reduced muscle bulk, loss of libido or the ability to sustain an erection.) Dopamine levels therefore need to drop, allowing prolactin levels to rise, in order to make more milk. Usually, this drop isn't noticed by breastfeeding mothers. In D-MER mothers, the dopamine drops too abruptly and causes the dysphoria. Given that D-MER has only had a name for about fifteen years, you can't help but wonder what sort of conversations women presenting with its symptoms used to have with their health-providers. Were the feelings chalked up to the nebulous 'baby blues'? I asked my own mother about this, who breastfed me in the early 1980s. She says she experienced some mild but noticeable dysphoria with me, my sister and my brother, but that she just thought it was 'part of the process'. 'You were told it was "just your hormones", as if that gave you any sense of clarity. You just said, "Oh, right" and got on with it.'

As for my own peculiar nipple sadness, various Google search terms led me to page after page after page of non-breastfeeding women saying the same thing happens to them. I wasn't expecting it, but it seems there are a lot of homesick-making nipples out there. However, I cannot find any research that really offers a proper explanation, save one fMRI study from 2011 by Dr Barry Komisaruk, a psychologist at Rutgers University, who has spent years researching the neuropsychology of orgasms and was the first to locate the female orgasm in the brain.[50] In 2011, Komisaruk recruited eleven healthy, non-pregnant women between the ages of twenty-three and fifty-six. While inside the brain scanner, each woman stimulated her clitoris, vagina, cervix and nipple by tapping rhythmically with a finger on the nipples and, for the vagina and cervix, using a plastic dildo. Again, you hope decent modesty curtains were positioned, or that they were playing one of Chopin's Nocturnes gently through the speakers. Nipple stimulation lit up the area of the brain that receives chest sensations, as expected, but a part of the brain called the medial paracentral lobule – right at the top, where the middle of a pair of over-ear headphones would be – where genital sensations are received, also lit up. Komisaruk hypothesised that stimulating the nipples, like in breast-feeding, releases oxytocin. As we know, this hormone triggers uterus contractions. So it's possible, he said, that nipple stimulation triggers contractions which then produce a sensation in the genital area of the brain. In other words, we can trigger a release of oxytocin *without* having a baby on the end of our nipple. As a woman yet to breastfeed a baby, I have to say I'm slightly nervous about how weird that sustained nipple action could make me feel.

Part Four

Part Four

'It's just chemical', or: what is fixed and what isn't

Given that female 'excess', however that might look, has always been linked to the reproductive organs, it would make sense that the discovery of sex hormones and women's changing biochemistry throughout her cycle – or any other reproductive process for that matter – might remove some of the stigma. That there could be a clear, chemical 'reason' for a certain type of behaviour or way of feeling is on the surface both legitimising and reassuring.

The idea really may be a profound relief to some. If any woman is able to ride out feeling low or extra sensitive for a few days each month by being aware of her cycle patterns; or finds telling herself that the way she's feeling is 'just chemical' is a strategy that stops her going up into that second gear of self-analysis and existential woe, I absolutely lift my hat to her. No, I doff it and curtsy. Personally, I am getting much better as time goes on at reminding myself that PMS is a transitory experience; that if I am sobbing into the door of the fridge or sinking into a bog of self-persecution over a pointless text argument with my partner which has escalated because a particular collection of words has hit a nerve, the feeling of not-quite-behaving-or-thinking-like-myself will pass.

This is not always easy in the moment.

When my hormone levels change, notably at ovulation and in the week before I get my period, it really can feel like I've stepped inside the skin of a different person. That I am not myself. But as I've dug more and more into the history of how women's bodies and minds have been positioned, and where that might leave us in the present day, a question has bobbed around my head: what *is* 'myself' if not a sensitive-by-design brain and body that is always changing and responding to the world it inhabits and the relationships with people around it? From what fixed, ideal state do I feel I am slipping when I am more emotional for the second half of the month?

What about me is fixed at all? If my thoughts and behaviours are so prone to changing when I'm in the run-up to my period, what does that tell me about the very *concept* of personality?

Personality

I have a personality. Or at least I think I do. Personality is interesting to consider when it comes to what is fixed and what isn't. When looking at how hormones affect who we think we are, I think it is particularly relevant. For decades psychologists have been fascinated with extracting the science of who we are by defining personality as the individual differences in the way people tend to behave, think and feel. Each person has some idea of their own personality type – whether they are sensitive or robust, shy or outgoing, etc. – but psychologists have largely focused on personality *traits*.

The most widely accepted are the Big Five personality traits, also known as the five-factor model or the OCEAN model. Developed in the 1970s, these broad dimensions used to describe the human psyche are: Openness (a person's openness to experience); Conscientiousness (a person's sense of organisation, discipline and duty); Extraversion (how assertive or buoyant a person is in social situations – introversion versus extraversion is the most recognisable trait of the Big Five); Agreeableness (the extent of a person's warmth and kindness); and Neuroticism (the degree to which a person experiences the world as distressing, threatening and unsafe). The model accounts for a person having various amounts of each. For example, I might have an average amount of openness, a lot of conscientiousness, a small amount of extraversion, buckets of agreeableness and plenty of neuroticism.

The Big Five model has always had its critics, mostly questioning how reliable the methods of testing are, given that it relies on self-reported data. The problem with people scoring themselves on tests is that we will often describe ourselves in a way that makes us seem more socially agreeable. This is commonly known as response-bias, and gives rise to many a caveat in the conclusions of number-driven psychological research. The Big Five model is also pretty limited in terms of how many traits have been identified. However, it has stood as a robust, universal human measure for decades and is still used widely in the corporate world to find the 'right' people for particular roles.

The most significant study that has cast doubt on the universality of the Big Five model and the evolution of personality was published in 2013 in the *Journal of Personality and Social Psychology*. Lead author Dr Michael Gurven, an anthropology professor at the University of California, had a team of researchers administer a translated version of a Big Five personality inventory to 632 Tsimane, members of a small tribe of hunter-gatherers in the tropical Bolivian lowlands.[51] The researchers asked them to rate on a 1-to-5 scale how much words like 'aloof', 'reserved' and 'energetic' described their personalities. Analysing the results, Gurven's team discovered that the traits did not group into the usual Big Five categories. For example, a person who rated himself as 'reserved' also said he was 'talkative', which suggests that the common conception of extraversion makes no sense to this culture. Only two groups of responses emerged from analysis of the forty-item test: industriousness and a tendency to be prosocial. Neither fit into the Big Five. It was a surprising finding, given that Dr Robert McCrae, a Big Five theorist, had found evidence to support the Big Five model in more than fifty countries. McCrae agreed at the time that the study raised important implications.

Consider, too, the Myers-Briggs Type Indicator (MBTI) test, used in countless industries and by many major organisations worldwide. The test was developed by two housewives, Katharine Cook Briggs and her daughter Isabel Briggs Myers, during World War II. They based it on the theories of Carl Jung, their aim being to create a useful test that would help place women entering the workforce in jobs that best matched their personalities. It is the most widely administered 'psychological' test in the world, estimated to be taken by up to five million people a year. Faith in the MBTI is still unbending. A *Forbes* article in 2014 estimated that the tool generates $20 million a year for its publishers.[52] Yet social scientists have been saying for decades that the MBTI has absolutely no evidential basis and is, largely, meaningless. It relies on binary choices (e.g. introversion versus extraversion), whereas no one is exclusively one thing or the other – human beings don't work that way; is wildly simplistic; and, like horoscopes, mostly sticks to positive language. None of the results will tell you that you're a simpering shit with no backbone. The MBTI also does not conform to the basic standards that are expected of psychological tests. It's quite fun to take, though, which is undoubtedly part of the problem.

Take the MBTI and you will be identified as one of sixteen personality types, referred to by a four-letter abbreviation. For example, INTJ ('The Mastermind'), which stands for Introvert, Intuitive, Thinker, and Judging. The trouble is, your MBTI can change over a matter of months, weeks, hours or minutes. I was an INTJ when I took the test at the beginning of my Psychology Masters in 2016. I took it again a few weeks later, as we were asked to do, and was an ESFP ('The Performer'): Extravert, Sensing, Feeling, Perception. I took it again while writing this book and was an INFJ ('The Counsellor'): Introvert, Intuitive, Feeling, Judging. Let's be honest, 'The Mastermind' sounds the best. But if a person's 'type' can change that quickly, four letters clearly offer very little insight into a person's character. However, the 'Eureka!' moment people may experience when they've taken the test (I felt it the first time around) speaks of something important: how easily seduced we are by the idea of an ideal self.

We are a variable, highly complex and often unpredictable species. That we are complex and changeable by nature isn't something that always sits well with us. Like moths to a flame, we flock to easy answers and solutions. It's comfortable, makes us feel secure. The MBTI might seem to offer one of those easy answers, yet no binary test could *ever* capture the variability of human beings. However much we might want it to.

Softening plaster

I am interested in how much and in what ways the fluctuations in our hormone levels can affect our sense of self; what makes me *me*. Because it is the ways in which we analyse how we feel, all the self-surveillance, that brings so much torture. As you can see, the existing means by which we make sense of ourselves as a society are not fixed or universally agreed upon. The nature of the language we use to describe who we are isn't either. So how can our personality be considered a fixed thing?

In 1890 the Harvard psychologist William James first posited the theory that, after thirty, we don't really change much. In his text *The Principles of Psychology* he writes: 'In most of us, by the age of thirty, the character has set like plaster, and will never soften again.'[53] Again, the fixed-by-thirty idea has been very pervasive over the years. But can it really be true that we're unable to change who we are once we hit the doddery old age of thirty? Many people do find it harder to change their habits and ways of being in their thirties and there are many psychologists who will agree that, unlike our moods, which are transitory in nature, personality reaches a peak of stability after puberty and into our mid-twenties. So to a certain extent our personality *has* taken shape by the time we're thirty. However, given all we now know about brain plasticity – the scientific term for our brain's ability to change throughout our lives – the reality is more complex.

Who we are cannot solely be a product of age. We are a complicated combination of our genetics, our upbringing and our experiences throughout life. What we do, what happens to us and who we meet affect our personality. When we are young our developing brains are much more pliant and the experiences we have shape us more. As we grow up, the nurture and support of others might be essential to make us feel we are accessing our potential, but bad or upsetting experiences too can be what we call 'character-building'.

We *build* our characters over time.

The foundations are there from when we enter the world through our mother's bodies. Much of our framework is the product of experiences we have in our early years, but we continue to build around that framework our whole lives. Who we are as a person can be affected by major life events like having children, suffering bereavement or changing careers. If we have had a difficult past, finding love with someone who truly sees us can be incredibly healing and freeing, allowing us to realise our potential.

Keeping this idea of our changeable selves in mind, one of the most common things women say when they're suffering the symptoms that may come with hormonal fluctuations, like in the latter half of our cycles or during the menopause, is: *I don't feel like myself.* Whether or not changes in our hormone levels *actually* affect our personality is an entirely subjective concept and almost impossible to measure. A woman on day 25 of her cycle may say one thing, a gynae-cologist may say another, a psychiatrist might say something else and a critical psychologist – that is, a psychologist who challenges mainstream psychological principles – might say that the only valid measure is what a woman says at any given time. I tend to agree with this last opinion.

We do know that our hormones fluctuate, though. Throughout the month, as I outlined earlier in the book, levels of different hormones go up and down in order to pre-pare us for possible pregnancy. That this biological process is happening is unarguable. Compared with all the myths and ritual surrounding women's bodies and minds in the past, that we have biochemical *reasons* for changes in mood or behaviour seems better. It's something 'real', measurable and scientific to hang on to, not the conjecture of ignorant – in the truest sense of not having knowledge, rather than just pig-headed – men in positions of power over us. But is it really *better*?

If we suffer really badly with PMS each month, a doctor telling us that there could be something wrong in the way our chemistry is working, bracketing us off into a diagnostic category, is alluring and potentially reassuring. Is it inherently kinder or more safe-making, though, than someone telling us our wombs control our brains like a character in a video game?

Is it possible that, although the cultural mythology of the past may have been eclipsed by biomedical science, it still serves, even with the best intentions, to manage female excesses? Could it be that we are still directly attributing female distress and deviance to the reproductive body? I think so. The problem is that we *don't* have precise scientific explanations for the ways in which our fluctuating hormones can affect our thoughts, feelings and behaviour. We have ideas from research in this area, which is ongoing, but no definitive answers. That we could be 'diagnosed' with something like PMS, giving us a seemingly clear 'chemical' explanation for our distress, draws an important parallel with the 'chemical imbalance' theory of depression that has prevailed through time without any real scientific basis.

'Imbalance'

For a long time, we were told that an 'imbalance' of certain chemicals in that big, wrinkly organ of ours was the cause of mental illness. Scientists first hypothesised this idea in the

1960s, after the apparent success of drugs thought to alter the levels of these chemicals. These drugs became known as antidepressants, 'discovered' by accident in the 1950s.

Scientists at the Munsterlingen asylum in Switzerland were seeking a treatment for schizophrenia and found that a drug which modifies the brain's balance of the chemicals involved in controlling mood, pain and emotions caused episodes of euphoria in patients. This was not beneficial for those with schizophrenia. It was, however, helpful for those with depression. The initial trial in 1955 saw patients report increased energy and sociability. The drug, imipramine, was called a 'miracle cure'. Pharmaceutical companies rushed to develop rivals. These were all known as tricyclic antidepressants, named after their three-ring chemical structure.

Many people reported a relief of symptoms, but side-effects were rife: exhaustion, weight gain and, sometimes, fatal overdose. Scientists produced a new alternative: the selective serotonin reuptake inhibitor (SSRI), said to home in on the neurotransmitter serotonin. Neurotransmitters are chemicals that help relay signals from one part of the brain to another. Serotonin is believed to influence mood, libido, appetite, memory, sleep, social behaviours and learning. It was posited that those with depression or anxiety disorders may not have enough going around.

This new class of antidepressant was led by Prozac, entering the US market in 1987. Patients reported the same kind of relief as with the first generation of antidepressants, only with fewer side-effects. Far easier to prescribe, SSRIs made their parent companies billions in the coming decades. Of course, the very notion of a drug so transformative had its detractors. Over the years, critics said antidepressants were prescribed too often, that their function was still empirically unproved and their long-term effects unknown. Nevertheless, sales remained robust. In 1994, *Newsweek* said, 'Prozac has attained the familiarity of Kleenex and the social status of spring water.'[54]

It is curious that the popular theory of chemical imbalance continued – and continues – to saturate public understanding of mental distress like depression and anxiety, because no robust evidence to support it has ever materialised.

Historically, psychiatry has shifted focus in its quest to understand the mind and determine what is or isn't 'normal'. In the first half of the twentieth century – owing largely to Freud's influence – organic brain function almost vanished from psychiatry. Prozac repainted the landscape. The chemical-imbalance theory continues to grab us because taking a drug to 'correct' faulty chemistry seems instinctive. It's a politically attractive idea, too. Social scientists argue that those who perceive themselves to be ill are easier to manage than those who feel their distress is a result of societal injustice. But while the 'depression is just like any other illness' narrative may be helpful for some, for others it reinforces the idea that they're different from those who are 'well'. This leads us to question the diagnosis of mental 'disorders' altogether.

The pharmaceutical industry has a direct interest in shaping behaviours and emotions into various symptoms, to be sold back to consumers as disorders requiring medication. The *Diagnostic and Statistical Manual of Mental Disorders* (DSM-5) still holds considerable influence in psychiatry, yet categories like bipolar disorder, schizophrenia, anxiety and depression cannot be 'confirmed' with tests, the way diabetes – often used as a comparator for depression in the 'Well, you wouldn't not take insulin if you were diabetic, would you?' line of argument – can. But that doesn't quite work as an analogy because the borders between psychiatric diagnoses are fuzzy. People who are depressed or have a diagnosis of bipolar disorder also have many symptoms of anxiety, for example.

Today, robust evidence shows poverty and inequality to be the biggest cause of suffering on earth. We know that our psychological constitution is affected by trauma in our early lives, that our bodies remember emotional pain and remain chronically inflamed. However, governments appear to remain selectively dismissive of such findings. Funding bodies like the Medical Research Council (MRC) continue to throw millions at identifying biological causes of mental distress. There has been some success in identifying genes that increase susceptibility to certain problems, but far more research on environmental factors is needed.

Being aware of this imbalance in the very mechanisms by which we understand mental distress is important because the tacit suggestion is that mental illness must be viewed as interchangeable with physical illness to earn interest and funding. The idea that mental distress is 'real' because it's lurking in our brain chemistry and our genes can reinforce stigma around what it means to become mentally unwell – that when we suffer, we are inherently broken or wrong. The tendrils of stigma extend to those who choose to take antidepressants as part of their mental-health management, but how a person chooses to conceptualise their distress should only be judged by how useful it is for them at any given time. Stigmatising the use of antidepressants is grossly unkind and unhelpful.

Clinical trials are not always reliable. Studies with questionable methodology and small numbers of participants continue to generate clickable headlines and affect public consciousness. But although we don't know exactly how antidepressants work, we know they *do* work for many people, helping to reduce catastrophic thinking patterns, increase energy and improve mood. They can save lives at points of crisis. We can say this while also acknowledging what else was going on when a person decided to take medication; for example, how someone began to accept themselves during talking therapy or felt empowered with increased social activity will also effect a change in thinking. However, given that we have no tests for what is helping most for an individual, isn't it up to us what we attribute our improved state of mind to?

When I have written about this subject in the past and posted the articles on social media, my feeds have been full of people saying that chemical-imbalance theories are washing away; that 'everyone knows' mental illness has many complex factors. I'm not sure I agree. Many mental-health professionals, particularly those who analyse the ways politics and culture underpin mainstream theories and practice, questioning the absolutes of empirical research, believe that it's stuck. Dr Joanna Moncrieff is a critical psychiatrist who writes compellingly on how chemical-imbalance theories still affect how we perceive our distress. She tells me that, although she does not usually initiate the prescribing of antidepressants, she says that 'often it is too difficult' to recommend that people stop taking them. People have been convinced they have a chemical imbalance and need the drugs to correct it, and are therefore too nervous to stop. She does, however, ensure they are fully informed – something she calls a 'drug-centred approach'. Moncrieff encourages 'public conversation that highlights the things about modern life that make us unhappy and asks more complex questions about what we believe depression is'.

What is happening under the Conservative current government in Britain at the time of writing this book is a naked illustration of what happens when people are lonely and isolated. As welfare and public services are decimated, people are becoming more distressed. It is there in the (vast) data. Antidepressant prescriptions continue to rise in part because threadbare services cannot match demand. GPs often have no other immediate options for people when waiting lists for talking therapies are so long. In turn, increased demands on NHS therapists has led to them reporting increasing levels of distress.

The 'biopsychosocial model' is widely accepted in most areas of human study and most healthcare systems; it works from the position that our experiences and environment shape our mental health in vastly variable ways, alongside possible biological factors. Still, that our state of mind is 'made' by factors both within and outside us isn't always presented in an accessible, mainstream way. There is evidence to suggest that framing mental distress as a biological condition can reduce a person's agency and make them feel less able to overcome their symptoms. We divorce the way we are feeling from a sense of self, othering the experience. *It's not me, Eleanor, experiencing depression; it's just that my brain chemicals have gone wrong.*

Any claim that depression is purely a result of reduced serotonin levels should be rigorously challenged. It does not fit with the evidence. This does not mean that medication affecting levels of serotonin in the brain, like SSRIs, absolutely don't work. It means that we just don't know if they're having an effect on the root cause of someone's depression.

Interestingly, SSRIs are a common treatment option for women who experience severe PMS. Has what we know about how hormones affect serotonin always been more of a precise science? Is the interplay of hormones and neurotransmitters the reason why, in the United States, antidepressants have been overwhelmingly marketed to women? Absolutely not. So much of the pivotal research in this area is recent. Still, in the twenty-one years since the US Food and Drug Administration allowed pharmaceutical companies to advertise prescription-only medications directly to consumers, the ads have always been *full* of women.

In a compelling *Huffington Post* piece ('Bad Mothers and Single Women: A Look Back at Antidepressant Advertisements'[55]) from 2012, the writer Katherine Sharpe presents a rogue's gallery of ads, published in major publications like *Time*, for antidepressants and anti-anxiety drugs from the 1960s to the present day. The collection shows how much the narrative around mental illness has changed, moving from an emphasis on character flaws to chemical imbalance. However, so much has stayed the same. From beginning to end the ads focus on family values and courtship, which Sharpe writes 'show women who need medication because they fail to thrive in female roles (lover, wife, mother), or because they are oppressed by the demands of those roles (the trapped housewife, the harried working parent) . . . the singletons, the wallflowers, the ineffective moms'. On one ad for Effexor XR (venlafaxine), an SSRI, the words 'I got my mommy back' are scrawled across the page in massive, multicoloured 'kid's' writing. It made me shriek. Sharpe, who wrote the critically acclaimed book *Coming of Age on Zoloft: How Antidepressants Cheered Us Up, Let Us Down, and Changed Who We Are* (2012), suggests that advertising 'must be at least partly responsible for the fact that over twice as many women as men use antidepressants' and it's hard to disagree.

The US and New Zealand are the only developed countries that permit this kind of direct consumer advertising from pharmaceutical companies. But that is a lot of potential insecurities – a woman's fear of failing as a mother, for example – being exploited. Under the shadow of Big Pharma, every anxiety can become a diagnosable mental disorder for which a product, a pill, is the solution.

In our world of expanding diagnostic borders when it comes to our mental health, along with the exhaustive range of psychotropic medication options, women's distress is still being positioned in the Western world as an excess that needs to be restrained.

PMS and neurotransmitters

The two best-studied neurotransmitter systems implicated in the psychological symptoms of PMS are those involving GABA and serotonin.

In the National Institute for Health and Care Excellence (NICE) Clinical Knowledge Summary (CKS), the prescribing of SSRIs is recommended for women with severe PMS. The guidelines state: 'For women with severe PMS (that is, women whose symptoms cause withdrawal from social and professional activities and prevent normal functioning), offer lifestyle advice and consider a selective serotonin reuptake inhibitor (SSRI, off-label use). The SSRI can be taken either continuously or just during the luteal phase (for example days 15–28 of the menstrual cycle, depending on its length).'[56]

The use of SSRIs in the clinical management of PMS is based on studies that have shown there may be less serotonin going around in women who really suffer with the psychological symptoms of PMS each month. However, most studies carry the caveat that the precise cause of PMS still remains unknown. There is a considerable body of evidence for the beneficial effects of SSRIs in treating PMS and, again, I always approach this subject with the 'whatever works' clause: if taking antidepressants is part of someone's mental health management because they've made an informed decision to do so, that decision must be respected.

I needed to talk to someone involved with the research into PMS to try to get some clarity, because this overarching mystery was starting to make me feel like I was on a teapots ride, spinning, every so often being able to make out something clear, only to start spinning again. I made contact with Mr Nick Panay, a consultant gynaecologist and the Chairman of the National Association for Premenstrual Syndrome (NAPS). Mr Panay is also the director of the West London Menopause & PMS Centre at Queen Charlotte's and Chelsea & Westminster hospitals, as well as heading a clinical research team which publishes widely and trains health professionals at all levels. I saw Mr Panay's name pop up on lots of online message boards, heralded as some sort of hormone god by women with severe premenstrual or menopausal distress who had been treated by him. I visited Mr Panay, a warm, brown-eyed man, in his Harley Street clinic. We chatted over an enormous wooden desk in a room that was almost certainly bigger than my entire flat. 'It's true,' he says, 'that we don't know exactly what causes PMS, but cyclical ovarian activity and the effect of estradiol and progesterone on the neurotransmitters serotonin and gamma-aminobutyric acid [GABA] appear to be key factors.'

We know about serotonin, but what's GABA? Not the sub-genre of hardcore techno music, no. That's gabber.

GABA is made in brain cells from glutamate, the most prominent neurotransmitter in the body, especially in the cerebral cortex, located in the super-wrinkly, outer layer of the brain. This area is responsible for processing information from the five senses, as well as higher thought processes like speech and decision-making. GABA is an amino acid that inhibits nerve transmission in the brain, calming nervous activity. Glutamate acts as an excitatory neurotransmitter and, when bound to adjacent cells, encourages them to 'fire' and send a nerve impulse. GABA does the opposite and tells the cells not to 'fire'. It is effectively our natural tranquilliser. Drugs such as benzodiazepines (diazepam, Xanax, etc.) work by increasing or imitating GABA's effect. As a health supplement, GABA is sold and promoted as a natural tranquilliser. (As you might expect, there is *some* evidence in support of GABA food supplements producing a calming effect, else they wouldn't have been made in the first place, but most of the evidence appears to have been reported by researchers with potential conflicts of interest.) It is also popular among body builders because there is some evidence to suggest that GABA increases human growth hormone (HGH) levels. There is evidence to suggest that low levels of GABA are associated with postpartum depression, although more research is needed.

Without GABA, nerve cells can fire too readily, too often. There is robust evidence to show that low GABA activity is associated with high levels of anxiety. We can use the example of caffeine to help understand the effect of GABA. Caffeine is known to rapidly increase glutamate levels, contributing to our alertness. GABA doesn't get a chance to get in and do its job. Remember, the less GABA there is, the more excitable our brain is. Now, remember the last time you drank too much coffee and how you felt. That awful feeling is, in crude terms, the sensation of glutamate in cruise control without the GABA brake pedal.

I am remembering when I worked at VICE and bought a fancy iced coffee from a new stall that appeared out the front one day. I have iced coffees a lot in the summer and they don't usually make me feel jittery or strange. This time was different. I downed the whole thing in about five minutes mid-afternoon and quickly felt like my scalp was peeling back off my skull. I couldn't see my computer screen and felt like my arms and legs were completely empty; no bones, muscle, fat, sinew, nothing. Empty tubes of skin. I felt sick, sweaty and scared, particularly that something profane was about to happen in my small bowel before I could get to a toilet. So shaky and paranoid was I in such a short space of time that I went out and asked the vendor if they'd done anything 'special' to it. 'You asked for a double, right?' the guy said. I said yes, meaning the usual double shot of espresso that would go into a latte or flat white. 'Ah, shit,' he said. 'I thought you meant a double *double* shot.' So I'd had an iced coffee with four shots of espresso in. Never, ever again. My gut twitches at the thought.

So, fluctuations in our reproductive hormones seem to affect the ways signals pass through our brain. 'Hormonal transitions like the fluctuations that happen across the menstrual cycle, postpartum, perimenopause and menopause seem to predispose some women to mood disturbances,' says Panay. 'The cognitive and behavioural symptoms associated with hormonal changes, what we know as PMS, usually start in the adolescent years when the fluctuations begin.'

I dug into the research again after our conversation and found a large meta-analysis (when data from multiple studies

144

is combined to identify common effects) from 2015[57] conducted at the University of Leipzig in Germany on the neuroscience of the adult female brain during hormonal transition periods. (Incidentally, the university is where, in 1879, the German physician and philosopher Wilhelm Wundt founded the first formal laboratory for psychological research.) Anyway, this review, the most exhaustive I could find, opens by saying that sex hormones have been 'implicated' in 'important mechanisms of neural plasticity', i.e. changes in the brain. We know this. But to what end? Does the very fact of having sex-hormone-producing ovaries leave us vulnerable to mental distress? Possibly.

According to this review, which homes in on depression, vulnerability for a period of depressive illness corresponds to the main hormonal transitions across the female lifespan. During childhood (0–9 years), a phase associated with low oestrogen levels in the blood, the prevalence rate for depression ranges between 2 and 3 per cent. When oestrogen levels start rising in puberty (10–15 years), so does the prevalence rate for depression, up to 8 per cent. During reproductive years, a phase when oestrogen and progesterone levels peak, prevalence rates vary between 21 and 38 per cent. Oestrogen and progesterone levels start declining during perimenopause (41–51 years), drop considerably after the menopause (45–65 years) and remain fairly stable during old age (above 65 years). This drop in sex steroid levels is paralleled by a decrease in prevalence rates for depression from 23 to 26 per cent during the hormonal transition phases to rates of 1–5 per cent during old age.

This gives us an interesting, if quite sad-making picture, but we must remember that not all women suffer with PMS. Not all women become depressed or consider themselves to be managing any other kind of mental health issue. It appears that some women are more sensitive to the effects of their fluctuating hormones than others. I will address why that may be in the next chapter.

Gender differences in mental health: not that simple

Everyone on this earth has their own set of neuroses. We all know what it's like to feel desperate, low, anxious or apathetic: these are facts of life when you have a human brain inside your skull. Changes in mood, even very significant ones, are normal. There are gender differences in common rates of mental health issues, but it would be both reductive and just not accurate to say that women are inherently more sad or mad than men.

As the World Health Organisation (WHO) advise in their guidelines on the prevalence of mental health issues across the world, these gender differences are complex: 'Depression, anxiety, somatic symptoms and high rates of comorbidity are significantly related to interconnected and co-occurrent risk factors such as gender-based roles, stressors and negative life experiences and events.'[58] Furthermore, we must remember that gender-specific risk factors for common mental health issues disproportionately affecting women include 'gender-based violence, socioeconomic disadvantage, low-income and income inequality, low or subordinate social status and rank and unremitting responsibility for the care of others'.

There is a significant gender bias in the treatment of psychological disorders. Doctors are 'more likely to diagnose depression in women compared with men, even when they have similar scores on standardised measures of depression or present with identical symptoms', and it is well documented that female gender is a significant predictor of being prescribed psychotropic (mood-altering) drugs like antidepressants or antipsychotics.

Gender differences exist in the patterns of seeking professional help for a mental health issue. Women are more likely to seek help from and disclose mental health problems to their primary healthcare physician, while men are more likely to seek specialist mental health care and are the principal users of inpatient care – that stiff-upper-lip-ness of masculinity and what it 'means' to be emotionally vulnerable often leads men to try to suppress their feelings, so they don't seek help until they become very unwell. We also know that suicide is the biggest killer of men under forty-five in the UK. In 2015, 75 per cent of all UK suicides were male. Men are more likely than women to disclose problems with alcohol use to their healthcare provider. 'Gender stereotypes regarding proneness to emotional problems in women and alcohol problems in men, appear to reinforce social stigma and constrain help seeking along stereotypical lines,' continue the WHO guidelines. 'They are a barrier to the accurate identification and treatment of psychological disorder.'

You can see how, in the case of a woman feeling debilitated with PMS each month, our rich, rich history of being thought of as inherently unstable might make her feel wary of seeking help in case she is dismissed with an 'Oh, it can't be that bad', 'Just get on with it', or 'Well, that's just part and parcel of being a woman, deary!' Sadly, this does still happen to women all the time. Although much progress has been made, the stigma of women's emotional excess still covers the surfaces of the medical establishment like a fine dust. I return to the question I posed earlier about which is better in helping us understand and treat the psychological symptoms that come with hormonal fluctuations: myth and ritual, or biochemical reasons and scientific 'proof'.

My instinct is, of course, to say the latter. However, although brain-imaging studies may have begun to shed light on the complex brain circuitry involved with things like PMS, postpartum depression, perimenopause and menopause, there is another question scratching at my mind: is it not all a bit … pathologising? Why do we need biological 'proof' for something like PMS when women have always known it exists? Why isn't how we say we feel reliable enough? Are the diagnoses themselves reliable enough?

In the 2011 study *Hormone-Specific Psychiatric Disorders: Do They Exist?* author Margaret Altemus wrote:

> There are several challenges to identifying hormone-related syndromes. First, in naturalistic reproductive hormone fluxes, such as puberty, the menstrual cycle, pregnancy, lactation and menopause, multiple hormonal changes occur simultaneously.[59]

In other words, with so much going on in a woman's physiology at any given time, it's extremely hard to be specific. She continues:

> There is an unfortunate tendency to attribute psychiatric symptoms to fluctuations in oestrogen, rather than considering a more complete set of hormonal changes. We cannot easily isolate one hormone, one neurotransmitter, one organ, or one area of the brain and blame that entirely.

In view of this, should premenstrual symptoms, even at the severe end of the spectrum, be considered a mental disorder?

Formal recognition

Much debate surrounded the inclusion of premenstrual dysphoric disorder (PMDD), a severe form of PMS, as a diagnosable psychiatric disorder in the 2013 fifth edition of the *Diagnostic and Statistical Manual of Mental Disorders* (*DSM-V*) – the diagnostic tome published by the American Psychiatric Association. It is used across the world as an authoritative guide to the diagnosis of mental disorders. However, no one I've ever met that works in psychiatry, psychology or any other mental-health setting in the UK adheres to its guidelines unyieldingly. In fact, many are highly sceptical of the *DSM-V*'s expansion: it now has a whopping 297 diagnosable disorders within its pages. If we are considering the possibility that women's normal variations in mood are still being pathologised, in a way that might not look like the worst bits of our history but nevertheless has the aura of something unhelpful, then it's important to think more generally about what the *DSM* is and represents in its current form.

The *DSM*'s latest expansion (shy? 'avoidant personality disorder' is now a thing) has pecked away more than ever before at the parameters of normal thoughts, behaviours and feelings that, even if a bit odd-seeming, never really used to be thought of as pathological. In other cultures they probably wouldn't be at all. Some of the additions are entirely medical in nature. For example, the 'disorder' 'caffeine intoxication'. To meet the diagnostic criteria someone must have five or more of the listed symptoms after drinking too much caffeine – be that three cups of coffee, cans of Red Bull or nosebleedingly expensive matcha lattes – which include: nervousness, rapid heartbeat, insomnia, restlessness and gastrointestinal upset. These symptoms must be impairing a person's functioning in some way to be viewed as disordered. Along with the intoxication part, 'caffeine withdrawal' is also included as a diagnosis in the *DSM-V*, the main symptom being a withdrawal headache but also including anxiety, low mood, low energy, difficulty concentrating, fatigue, nausea and flu-like symptoms.

According to the *DSM*, it seems like there are very few people on this earth whose mental health is pretty good. Jesus, as I type this, my own feels pretty good yet if I start turning the pages of the thing I know I'll fit any floating neuroses into the diagnostic brackets of at least twenty conditions.

Another contentious point about the *DSM-V* is how much diagnostic thresholds have been lowered. In the *DSM-IV*, to meet the criteria for a diagnosis of 'generalised anxiety disorder' one needed to have three out of six symptoms of excessive worrying for at least six months. In the *DSM-V*, this has been reduced to one symptom for three months. If you are in the middle of a health crisis and have spent three months extremely worried about it, or, perhaps, are unemployed and anxiously looking for work, to the point where said worrying feels uncontainable, the specific context of your worrying would seem to be irrelevant: you may very well be considered to have a disorder. In the past, you might not have been.

Dr Panay told me that PMDD affects between 5 and 10 per cent of menstruating women, a figure corroborated by the Royal College of Obstetricians and Gynaecologists (RCOG). Online, the NHS website links to information provided by Mind, the mental health charity, which says: 'Premenstrual dysphoric disorder (PMDD) is a very severe form of premenstrual syndrome (PMS), which can cause many emotional and physical symptoms every month during the week or two before you start your period. It is sometimes referred to as "severe PMS".'

Medical literature was, until quite recently, vague about what PMDD is and how to treat it. One of the problems being that, from month to month, a woman's premenstrual symptoms may vary in length or severity. 'No two cycles are ever completely identical,' Panay tells me several times during our interview. He is one of the most prominent clinicians behind PMDD awareness in the UK and across Europe, but wants to make it clear that 'a woman's physiology and environment will never be the same month on month because we're just not built like that.'

Previous versions of the *DSM* chucked PMDD in the 'depressive disorder not otherwise specified' category. Things have changed. In the *DSM-V*, the PMDD diagnosis has three main criteria. First, a woman's symptoms have to correspond with the menstrual cycle for a minimum of two successive months. Secondly, the symptoms must be truly disruptive to a woman's ability to carry out her normal activities. Thirdly, to be diagnosed with PMDD a woman must report that she is not depressed or anxious all the time, only in the second half of her cycle, from ovulation until the beginning of the period.

The inclusion of PMDD as a diagnostic category was controversial to say the least. Some feminist groups criticised the inclusion as a move that pathologises women's normal hormonal fluctuations; one that runs the risk of people using the status of a recognised diagnosis to claim that women are less capable because their emotions are intrinsically unstable. Whenever women are direct, assertive or say they're finding

151

something challenging, the fear was that this 'legitimate' label could see the 'She's hormonal!' slur being thrown around more. Or, that it might influence the promoting of women to high-powered roles because they *might* just lose it each month.

Altemus, writing her study while the *DMS-V* was in preparation, questions the proposed inclusion of PMDD:':

A stated goal of the *DSM-V* process is to try to use the biological pathophysiology of mental disorders to inform psychiatric diagnoses, including dimensional features which may cut across diagnostic categories. At this point in time, however, biological markers have not been identified which are robust enough to be incorporated in diagnostic criteria.

Still, it's there.

Another noted critic of the *DSM-V*'s PMDD inclusion was Sarah Gehlert, Dean of the College of Social Work at the University of South Carolina. She wanted to find out how many women actually have PMDD and to see if any robust evidence base could be established for the disorder – one she feared could be used against women. 'Say a poor woman was in court, trying to see whether she could keep custody of her child,' Gehlert said in an interview with NPR. 'Her partner's or spouse's attorney might say, "Yes, your honour, but she has a mental disorder." And she might not get custody of her children.'[60]

Premenstrual distress being used in court is an interesting and, of course, controversial subject. Should PMS be seen as a mitigating circumstance when criminal activity has happened? A conundrum that goes right to the heart of the old

152

question: mad, bad or sad? I found a *New York Times* article[61] from November 1981 detailing two cases where it had been. On 9 November 1981, a 29-year-old barmaid from east London was put on probation for three years for carrying a knife and threatening to kill a police officer. Sandie Smith had nearly thirty convictions for offences like arson and assault and was already on probation. Her defence team argued that all these crimes had coincided with her premenstrual phase which, unless she was treated with the hormone drug proges-terone, rendered her 'a raging animal' each month.

On 10 November of the same year, a 37-year-old woman named Christine English pleaded guilty to manslaughter by reason of 'diminished responsibility' on an original murder charge. She was conditionally discharged for one year by a Norwich court. In December 1980, after an argument with her lover, English had driven her car into him. Her lawyer argued in court that she was experiencing 'an extremely aggravated form of premenstrual physical condition'. A medical witness corroborated, saying she had suffered from the condition since 1966. The judge, reads the article, was satisfied that English had committed the crime under 'wholly exceptional circumstances'. At the time, Dr Anthony Clare, a psychiatrist at the University of London's General Practice Research Unit, said any attempt to define PMS as a condi-tion was 'a nightmare', since there is 'no consistent biological abnormality'.

At both trials the crucial defence witness was Dr Katharina Dalton, who had been treating English. Dalton was an eminent British gynaecologist who, beginning in the early 1950s, challenged the widely accepted view that PMS was all in women's heads. People had a hard time believing it was real. Dalton led pioneering studies, overseeing one of the

153

first clinics to specifically identify and treat the symptoms of PMS, at London's University College Hospital. She developed the use of menstrual cycle charting for the diagnosis of PMS, arguing that the timing within the cycle directly correlated with higher risk of alcohol and substance abuse, violent crimes and suicide attempts. Dalton's dogged advocacy of progesterone as a treatment for PMS, arising from her theory that symptoms were the result of progesterone deficiency, has not been widely supported. What has been is her later work that emphasised how PMS could exacerbate pre-existing mental health issues like depression. Most experts today agree that women with a propensity for low moods or significant anxiety may find their symptoms are worse around the premenstrual phase. Dalton's testifying in defence of Smith and English remained controversial after her death in 2004. It was argued that, in appearing in those courts, she was perpetuating the old monstrous woman myths; was asserting that a natural biological process had made these women mad and out of control. Culpability and responsibility were not relevant.

A case with parallels to those 1981 trials took an interesting turn very recently. In August 2018 the Rajasthan High Court in India acquitted a woman accused of murdering a child more than three decades ago. This was on the grounds that she was 'suffering from insanity' triggered by premenstrual stress syndrome (PMS) at the time of the crime. Kumari Chandra was accused of pushing three children, two boys and a girl, into a well in August 1981. One boy and the girl were saved by witnesses. The other boy had drowned. 'In the present case, not one but three doctors, who treated her on different occasions, have deposed in favour of such plea of insanity set up by the defence,' a copy of the court judgment read. Chandra's lawyer, a man named Vivek Raj Singh Bajwa, argued that the trial court was wrong in convicting her for kidnapping the children with intent to kill them. Bajwa argued that she was suffering from a 'mental disease' known as premenstrual stress syndrome, which made her 'dangerously aggressive' in the days before her period. His defence relied on citing reports by Chandra's treating doctors, one of whom claimed she had such severe PMS symptoms he had to treat her with tranquillisers.

It wasn't *their* fault, see. It was their hormones.

You can understand why women like Gehlert would be wary of premenstrual distress being counted as a disorder, by definition a word that means a state of confusion, chaos, disarray. In her study into PMDD, Gehlert's research team randomly recruited 1,246 women and asked them to fill out a form every day for two months that asked simple questions about their mood. It said nothing at all about menstruation. 'I wanted to go into it as scientifically and objectively as possible,' Gehlert said. The women supplied daily urine samples so the team could identify where they were in their menstrual cycles. When the data were analysed at the end of the study, just 1.3 per cent of participants fitted the diagnostic criteria for PMDD. This is a much smaller number than other researchers have found for PMDD. Gehlert remained unconvinced. Referring to how little hard evidence we have for how hormonal fluctuations interact with the brain processes responsible for our emotions, she said, 'I would feel much, much more comfortable if we understood the biology behind it. Even though we found evidence, the question remains: Is what we described real?'

Another concern about premenstrual distress being conceptualised as a 'disorder' is how ripe it is for money-making.

Take the prescription drug Sarafem, approved by the Food and Drug Administration in 2000 as a treatment for PMDD. This drug is chemically identical to the SSRI Prozac (fluoxetine). The global pharmaceutical manufacturer Lilly faced losing money to generics when the patent on Prozac was about to expire, so decided to give it a new name: Sarafem. These pink-and-purple capsules come cased in a lovely pink packet with a sunflower on the front. A generic SSRI that cost 25 cents a pill was now being marketed as a PMDD-specific drug. Its cost? $10 a pill.

The marketing of Sarafem was a talking point. I managed to find clips of the original commercials online and found myself absolutely bellowing. Bear in mind this was only nineteen years ago. In one TV advert, a woman is seen trying to shove a shopping trolley back into a line outside a supermarket. She rams and grimaces, rams and grimaces. The narrator's voice (female) comes in: 'Think about the week before your period. Do you feel irritability? Tension? Tiredness? Think it's PMS? Think again. It could be PMDD.' In another advert, a cross woman in a suit yells upstairs to her invisible partner, 'Did you take my keys?' The narrator (female, again) pipes up: 'Sound familiar? Why put up with this another week?' Then it gets a bit *Terminator*-y. 'It's back. The week before your period.'

The FDA sent Lilly a cease-and-desist letter for trivialising the seriousness of PMDD. The 'crazed' women in the ads just look like normal women having a shitty day. Having a shitty day would add frustration, sadness and anxiety to any woman in the premenstrual phase – if she has symptoms, that is. As Gehlert said in her interview with NPR, this kind of representation could make people think 'that women – over men – were predisposed toward that sort of behaviour'.

Validation

In understanding how easily the PMDD diagnosis can feed into harmful stereotypes about women, what we must not forget is that formal recognition – or diagnosis – can make a lot of difference to women who are suffering. Women's pain – physical or emotional; as if the two can be extricated – has been minimised and dismissed for centuries. Many women diagnosed with PMDD feel their pain has in the past been invalidated and so its inclusion as a diagnostic category proper is more than welcome; it's being seen. Women suffering in this way may too have a better chance of getting the full attention a full diagnostic status presents.

Hannah Ewens, a 26-year-old journalist in London, has experienced extreme distress before her period since the age of fifteen. Her journey has involved several 'completely bewildered' GPs. After it was suspected that she might have a bipolar disorder, a link between her mood fluctuations and her cycle was established by a psychologist, though many years of suffering ensued. 'I tried everything,' she tells me. 'Changing [contraceptive] pills, not taking the pill, supplements, different antidepressants, everything. I'd spend at least three days suicidal every month.' Eventually, Ewens demanded her GP refer her to a gynaecologist. He resisted, but eventually acquiesced: 'I knew my rights and was armed with printouts of my own research on PMDD.' Since January 2016, as a result of that referral and the dual treatment of transdermal (applied to the skin) oestrogen gel and a Mirena coil that releases small amounts of progesterone, completely suppressing her cycle, Ewen has seen a phenomenal improvement in her mental health. 'I still get anxiety,' she says. 'But nothing like before.'

Ewens's story is extreme. That initial perplexity she met with at primary care level, though, will ring true for many women. There is a growing online community around PMDD now and plenty of forums where women can talk with each other about what has worked for them in alleviating their symptoms. Sometimes it's contraceptive pills or other hormonal treatments. Sometimes it's antidepressants. Sometimes it's diet and exercise. Sometimes it's psychological therapy like cognitive behaviour therapy (CBT). Sometimes it's a combination of all these things. Women can find each other on Twitter (with the #PMDD hashtag), Facebook and other social media networks. The space they provide for open discussion is a kind of refuge. In the UK, the National Association for Premenstrual Syndrome (NAPS) is a major resource.

Laura Murphy has set up a patient-led awareness project called Vicious Cycle: Making PMDD Visible. There is both a Facebook page (with nearly 10,000 followers) where women can communicate with each other and a formal website, which Murphy describes as 'a grass-roots project, passionate about raising the standards of care for sufferers of PMDD and extreme PMS'. With it, she says, they aim to 'raise awareness of PMDD among healthcare professionals and the general public' and to arm sufferers with 'the knowledge and tools they need to get treatment and support'. (My usually right-on female GP said 'What's that?' when I referenced PMDD in a consultation with her about what I could do for my own PMS woe.) Over Skype, Murphy, who is thirty-nine, told me about her experiences at length. She is a very articulate and funny woman who has, with her campaigning, become a member of the RCOG's Women's Voices Involvement Panel. I can see why.

'My problems started when I was about seventeen and was put on the [contraceptive] pill to help with my heavy periods,' she tells me. 'But I had my first ever panic attack, during which it genuinely felt like the world was ending, on day 21. I fell into a horrible depression, struggling to get up in the morning, and ended up dropping out of school.' By her mid-twenties, Murphy's premenstrual symptoms were so debilitating that she ended up sleeping 'up to eighteen hours a day' and flying into rages right before her period. She had frequent suicidal thoughts and said her ex-partner told her it was like 'living with a totally different person' every three weeks. When she was thirty-two she was advised to have a Mirena coil fitted. For some women, the progesterone-releasing coil inhibits ovulation and they experience a reduction in the heaviness and duration of their periods, along with reduced symptoms of PMS. After a couple of years feeling quite well with the Mirena, Murphy says her mental health deteriorated again. 'For eighteen months I was severely depressed and began experiencing awful anxiety for the first time in my life. My GP said that the Mirena couldn't be responsible because the hormones are only released locally. They prescribed me a drug called pregabalin [a medication used to treat epilepsy, neuro-pathic pain, fibromyalgia and, increasingly, generalised anxiety disorder] but I had suicidal thoughts for months on end. My hair started falling out. I couldn't work. I started having therapy and my counsellor was amazing.' She seems exhausted recounting it. 'It *was* exhausting,' she says. 'When you have sat in front of so many doctors who have told you, for the thousandth time, that it was "just" PMS and some-thing that all women have to deal with in life, you start to question your own sanity.'

After one of said doctors' appointments, Murphy went home and googled 'severe PMS' and came across the term PMDD. She says it was 'life-changing'. 'It's me!' she thought; the exhaustion, the depression, the sleepiness, the suicidal thoughts, the adverse reaction to the various methods of birth control. 'It was a total lightbulb moment.' She smiles. She recounts the day she almost 'walked in front of a lorry' and went to see her GP again. Thankfully, this GP referred her to a PMS and menopause clinic she had heard about on a PMDD Facebook group – the one led by Mr Panay. Murphy says that sitting opposite him in the clinic was the first time in her life she had sat in front of a medical professional who understood what she was describing. 'I was thirty-four years old and had spent half my life without the right diagnosis,' she says. 'It was more than a week in every month that I felt the way I did and it felt like it was getting worse and worse as time went on.'

Panay tried some different treatments with Murphy, including GnRH agonists, a group of drugs that are modified versions of the gonadotropin-releasing hormone (GRH), which helps to control the menstrual cycle. When used for two weeks or more the drugs stop the production of oestrogen, which effectively inhibits the cycle. The treatments didn't work for Murphy – in fact, they made her feel worse. In October 2017, she elected to undergo a full hysterectomy with bilateral oophorectomy (removal of the ovaries) – the last option in treating PMDD. As Panay tells me, it is the only option that could be considered a definitive 'cure', because it removes the very mechanisms that cause hormonal fluctuations. 'It was a huge decision to make, obviously, but in the end I felt like it was my only option,' Murphy says, matter-of-factly. 'I feel like I am beginning my life again.' HRT is often offered when the ovaries are removed, to replace some of the hormones they used to produce, reducing any menopausal symptoms as the body goes into immediate menopause after the operation.

Murphy's story is not uncommon. Within the forums she talks about, women discuss belittling-GP horror stories and share any insights they pick up along the way. We talk about the implications of a hysterectomy and how she has spoken to women who have struggled to convince doctors that it might be a worthwhile option for them, because the doctors just cannot fathom how they'd want to remove their ability to get pregnant. 'Obviously that was part of my consultation process,' says Murphy, 'but I was very clear that carrying a baby would not be part of my life, but another private consultant I saw told me he was not happy doing the surgery as I had not "had any children yet" – without even asking if I even wanted to have children.'

In one forum I read about a woman, already mother to a young boy, who had been suffering with PMDD since she was a teenager, who says she got on her hands and knees in her doctor's office and begged for the operation after he said, over and over, 'Well, you can't be sure.' The more I think about this the more it bothers me. It goes right back to the woman-as-birthing-machine edict. Why would a woman even begin considering putting herself through an operation as big and definitive as a hysterectomy, with its painful recovery period, if she wasn't in hell every month? Why should her physical ability to carry a pregnancy eclipse her state of mind and overall health, her ability to realise her potential or have happy, fulfilling relationships? Why does our fecundity trump everything? Why cannot a woman be trusted not to change her mind?

As Murphy is keen to point out, 'there is no one-size-fits-all' treatment for this hormone-based mood disorder. Panay explains something that seems clear from the minute you start reading women's stories: different things work for different people and, very often, you won't know what will work until you try. 'Sometimes women find enormous relief with the combined contraceptive pill,' he says. 'Sometimes, it makes them feel even worse. What we need to focus on is working with healthcare professionals and help inform women about what options they have.'

Truth serum

Clearly, if a woman's premenstrual distress is such that she feels unable to function, she should, in an ideal world, be seeking and receiving dedicated healthcare and support. I wonder, though, if there is something wider we should be asking about how, as a society, we frame what women say and do when they are anything other than sanguine, nurturing and polite? What if, as women, we could think about these emotional changes we experience in a different way? With PMS, could it be that we are saving up anger, frustration, our base need for affection, our tears and our ever-simmering sense of injustice for three weeks of the month? Then, as our hormonal levels shift, could they be acting as a kind of 'truth serum', lubricating the passage of what we *really* want to say or do? It's a radical idea and I quite like it. What if what we feel, say and do during those times of the month when we are so quick to say we are being irrational, needy or defensive is the most 'real' we are? Because the thing is that the science surrounding whatever signalling changes happen in the emotion-processing parts of our brain during the menstrual cycle is still imprecise. Could it be that, when the mantle of self-censorship drops and we are less caught up with fearing how we'll be seen, we're accessing all that historical oppression and, in short bursts, letting it go? All that so-called 'excess' could just be truth.

We need to try to do better as women to remind ourselves of the changeable beings we are by design; to accept that it is literally impossible to be a certain way all the time. We are not *supposed* to stay the same. Moods, too, come and go. Anger can be a force for necessary change in our lives. If we continue to root the entire blueprint of our personality, what makes us *us*, in our reproductive system, we're being neither kind nor smart.

Diagnosis: not for everyone

Robust evidence tells us that having an official mental health diagnosis doesn't help everyone. It's not just the language that matters, either, it's the systematic effect and meaning of that diagnosis. I have lived with a propensity for anxiety and panic attacks since I was a teenager. I have taken medication and continue to have therapy. There is a clear hormonal aspect to when my anxiety levels are higher. Taking all these things into account, I have always intrinsically rejected the idea of being 'diagnosed', because although my fight-or-flight response may run on overdrive and although I can slip into days of existential torpor sometimes, there is something in me that kicks against the idea that I am *dis*ordered. How can something as subjective and labyrinthine as a mind be labelled as, well, wrong?

These are informed personal feelings. Everyone is entitled

to their own and there is no right or wrong response. Sometimes, a diagnosis or label can make someone feel a lot safer in themselves; the terms are universal touchstones for uncomfortable ways of feeling. The idea that there is a name for your distress, rather than it be coming from 'nowhere' or be something totally out of your control, can be very comforting. Having a diagnosis also means being able to access the associated support and benefits that come with formal identification, and a common connection with others who feel the same. There is great power in this.

An important paper was published in the *Lancet* in 2018 reporting the results of a wide-ranging review of service user, carer and clinician experiences of mental health diagnosis.[62] For some people, psychiatric diagnosis was useful, and any problems stemmed from it not coming early enough. For others, a diagnosis was very oppressive. There is often a tension that exists between these two camps of people, particularly on social media, so it always bears repeating: any psychiatric diagnosis is not a single, defined thing. This paper suggests that some diagnoses are more helpful to people than others. A diagnosis of obsessive-compulsive disorder (OCD) or depression, for example, is likely to validate suffering and give the person a platform from which to access help and talk about their distress. There is not the same stigma attached to these diagnoses as to others associated with more serious mental illness, like bipolar disorder or schizophrenia. In a piece for the *Guardian*, the clinical psychologist, writer and active campaigner Dr Jay Watts wrote: 'With the latter [i.e. schizophrenia], diagnosis can produce what the philosopher Miranda Fricker has called "testimonial injustice", an inbuilt prejudice that gives less credibility to the diagnosis.'[63]

Borderline personality disorder: a bad name for real suffering

One of the diagnoses that comes with great stigma is borderline personality disorder (BPD) or emotionally unstable personality disorder (EUPD). I wanted to talk about this because it is a highly divisive concept in psychiatry and psychology, thought by some practitioners to be a modern-day version of 'hysteria'. Not least because three out of four BPD diagnoses are directed at women.

BPD has been constructed as a 'disorder' that is characterised by emotional turbulence or unpredictability, anger, unstable relationships and profound fear of abandonment. However, as Dr Jay Watts pointed out in another piece for the *Huffington Post*, 'BPD has always been a synonym for the "difficult patient" in psychiatric-speak.' Patients are constructed as 'too sexual, too clever, too aware of their actions to deserve care, interest and respect', she continues. 'Angelina Jolie in *Girl, Interrupted*. Glen Close in *Fatal Attraction*. Wasting resources and messing with staff's heads deliberately.'

Semper femina.

Deliberate self-harm is common in people who have been diagnosed with BPD. However, when they are seeking treatment for self-inflicted harm, including drug overdoses, they have been shown in research to be seen as 'difficult', indulging in 'bad behaviour' or, heartbreakingly, a 'nuisance'. This is one of many reasons why Watts, along with many other prominent clinicians and a growing online community of those diagnosed with BPD refers to it as a 'dustbin diagnosis'. I spoke to Watts, a warm, raven-haired and raven-browed woman whose passion is immediate, about this in more detail. To her this diagnosis doesn't just belong in the bin because of its lack of construct validity, it belongs there because it 'shuts down our human response'. I completely agree. Language is incredibly powerful and (as discussed earlier) the nature of personality is highly subjective, so to put it on a very basic level: what gives one person a right to tell another person that their personality is disordered?

There is a gender bias in giving this diagnosis. Take a recent study from 2015, titled 'Judging a book by its cover',[64] in which 265 clinicians were asked to watch a video of a woman presenting with 'panic disorder'. After, they had to analyse her presentation and prognosis. The group were split into three. A third were given the label 'past PD'. Another third were given her presenting problems as well as a description that is consistent with 'BPD', The others were only given her presenting issues. Analysis revealed that the group who were told that the woman had 'PD' gave her a poor prognosis and described her distress in the video in worse terms.

Why does this matter? It matters because we construct our sense of self through how people react to us. This happens all the time as we go through life. If we are told we are 'needy', 'unpredictable', 'manipulative' or 'angry' all the time it is going to reinforce the negative relationship we have with ourselves.

The reason many clinicians reject the BPD diagnosis, and why there is such a growing network of 'survivors', is because evidence shows that around 80 per cent of people diagnosed with BPD – remember: overwhelmingly applied to women – have a history of trauma. When clinicians like Watts are passionate about informing and empowering people, information travels and people may feel they don't need to take the oppressive labels stuck on them at face value. This is why

166

the #TraumaNotBPD hashtag has gained such momentum on social media. People want to reclaim power they feel they have lost not just in their past, but again in their present.

Trauma survivors may have a block on memories of what happened to them in their early life, but the impact of physical, emotional or sexual abuse as a young person can be profound and enduring. People who were sexually abused as a child often feel they are to blame. That they are somehow stained. On a subconscious level they believe they should be punished. The quick-to-fly-off-the-handle aspect of BPD, seemingly in relation to trivial things, can be conceptualised as a delayed expression of anger towards the person that caused them pain. Of course, not all trauma is physical or sexual in nature. Less-easy-to-define experiences like emotional neglect can also profoundly affect a person. It goes beyond the mind. In fact, study into the effects of trauma has strengthened what we already know about how deeply the mind and physical body are connected.

Trauma stains our fibres. Adverse Childhood Experiences studies (ACEs) show that childhood trauma, neglect and structural oppressions 'come out' in later life not just in the form of mental distress like anxiety or depression, but in chronic physical inflammation, bodies stuck in high-alert mode. Research has identified potential biomarkers for childhood emotional trauma, mostly focusing on the idea that abused or neglected children have higher levels of the stress hormone cortisol. In essence, they are waiting for trauma to recur. That stress hormones affect how our brains work, in terms of emotional responses or higher learning, is thought to be important. However, a systematic review of twenty-one studies in 2015 revealed that only between 0 and 22 per cent of psychiatric patients are asked about past trauma and violence. Given that it's proven to be rare for survivors to talk about past trauma without being asked, a culture of silence is perpetuated.

We know that research into women's hormonal sensitivities is in an early stage, but there is preliminary evidence to suggest that there may be a link between changing levels of ovarian hormones and BPD feature expression. In layman's terms, women with BPD might have worse psychological and behavioural symptoms of PMS. A consistent finding in longitudinal studies is that women with severe PMS have higher incidences of past physical and emotional abuse. It has also been shown that some women with PMDD have diminished positive emotional processing.

Women have been historically conditioned to keep their pain and emotional excess contained. We often don't feel we can speak because, when we do, we can be re-traumatised through damning language. The result is a body that constantly believes it needs to protect itself; over-alert and often in pain. Trauma moves through us with our blood. As the psychiatrist Bessel Van Der Kolk writes in *The Body Keeps the Score: Brain, Mind, and Body in the Healing of Trauma*, his now hugely influential book exploring the lifelong cost of burying traumatic experiences: 'As long as you keep secrets and suppress information, you are fundamentally at war with yourself.'[65]

The links between chronic pain disorders (like fibromyalgia, a condition affecting far more women than men and known to 'flare' in the premenstrual phase) and trauma have also been well established in recent research. Clearly, the entire system by which we identify and help people in distress needs to change. Thankfully it does appears to be changing. Slowly. Trauma Informed Care is a movement which is starting to grow internationally. The approach changes the focus of care from asking someone 'What is wrong with you?' to 'What has happened to you?' 'We must insist on trauma-informed care which focuses on listening to and helping women on their own terms, rather than attaching a label which provokes hate and disdain from even psychiatric staff and has led many a woman to suicide,' Watts, who is open about being a 'psychiatric patient' in the past, tells me.

The weight of trauma

Watts emphasises the fact that women who have traumatic experiences in their early lives often don't have a solid sense of themselves. In my Psychology MSc work placement with the clinical psychologist who both ran a chronic pain service and provided psychological assessments for people who had been referred for potential weight-loss surgery, I observed many assessments with women whose BMI put them in the 'morbidly obese' category and who, under the care of a bariatric surgeon, were hoping to have gastric bypass surgery to help them lose weight. The psychologist was responsible for assessing whether the patient was emotionally equipped to deal with such a major transition, and to get a sense of whether more intensive psychological support would be beneficial to them before and after the surgery.

I was shocked to learn that, for many of these women, in that room with the psychologist was the first time they had disclosed past sexual, physical or emotional abuse. Among the stories I listened to, as the (female, wonderful) psychologist listened and subtly made notes, I heard women describing childhoods spent in and out of care, sexual abuse within family settings, emotional neglect and violence. Many had mental health issues that they'd received patchy care for. I had not anticipated or considered there would be such a clear connection between historic trauma and disordered eating habits. But of course it makes sense.

According to the major Adverse Childhood Experiences (ACE) study published by the Centers for Disease Control and Prevention, more than six million obese and morbidly obese people are likely to have suffered physical, sexual and/or verbal abuse during their childhoods.[66] It's likely that millions more will point towards other types of childhood trauma as the cause of their weight issues: living with a mentally unwell family member, for example, or an alcoholic parent. A considerable body of research now shows that post-traumatic stress disorder (PTSD) is associated with an increased risk of women becoming obese. Yet as a society we still seem to blame the individual for their obesity. It is particularly pernicious where women are concerned. *Look at her! She's massive! Why doesn't she just get off the sofa and stop eating chips?* A big woman takes up too much space. *She doesn't look in control. She can't help herself.* How often do we think: what has happened to her?

Historically, fat was seen as the embodiment of social, economic and sexual wellbeing, but now the opposite is true. Body image ideals are social constructs born from the synthesis of history, politics, class and people's moral values and, today, modern Western culture celebrates thinness and denigrates excess weight. The media persistently reinforce the message that being fat is not only detrimental to a person's overall health, but is also viewed as irresponsible in both a social and economic context. An obese person's understanding of their physical presence and place in society is shaped by the societal discourse that surrounds them. They may experience a high level of shame, guilt and embarrassment in relation to their weight. This, in turn, may lead to decreased motivation to lose weight. It's all so wrapped up in class, too. We know that rates of obesity are higher in lower socioeconomic groups. Food choices are hugely influenced by income, knowledge and skill. It is always, always a much more complex picture than someone just not being able to stop eating burgers.

In the ACE study many participants described overeating as having benefits in their early lives. Binge-eating became a source of comfort and protection from sexual abuse. When overeating becomes a coping strategy for emotional distress at an early age, it is difficult to undo that conditioning as an adult. The reward is still the same. Distress can be momentarily squashed with food. Viewing overeating as driven merely by 'addiction' obscures the reality of the problem. Another connection between childhood sexual abuse and obesity might be a desire to 'de-sexualise' with excess weight, as a means of protecting against more abuse. Those who engage in binge-eating are in a feedback loop: they eat large amounts to feel better when feeling distressed, then, feeling disgusted with themselves for doing so, they purge. The subconscious need to emotionally soothe with food, yet also use it as a means by which to reach the jagged edges of shame and self-hatred, speaks of very deep, unexplored shame and pain. Once that shame, literally weighing the body down, begins to be opened and accepted, there is great potential for learning to re-route these destructive patterns. In that room where I observed women talking about sexual abuse to a professional for the first time in their lives, I could see the visible relief at being heard, but also the incredulity that their past experiences would be significant in this context. Remember, most of these women had a history of mental illness. Why weren't they asked about what had happened to them? We must do better at giving women a chance to speak their truth.

I thought about the women I met in the weight-loss surgery clinic when I was talking to Watts. 'So often for these women that are demonised there has not been someone around to tell them and show them, through the micro-interactions that form our very being, that they are "okay"', she says. 'Add into this a misogynistic, hetero-patriarchal society that, in turn, idealises and denigrates women, that pedestals us and yet also expects us to mould ourselves seamlessly to every different encounter, and no wonder women break down.'

We talk further about how women labelled with BPD are, in fact, having 'understandable reactions to impossible, contradictory pressures to achieve secure selfhood in a mad, mad world'. So rather than pathologise women with a discriminatory classification, one that 'character assassinates and gaslights' women, Watts believes we must see the construct of BPD as the misogynistic insult it is. A modern-day 'hysteria'.

Conversion

Sometimes, our bodies 'convert' emotional pain into something else. Something tangible. It was Freud who proposed that the memory of trauma which a person cannot confront because it will cause too much distress can be 'converted' into physical symptoms. Interestingly, such cases are often seen by neurologists today and often referred to as functional neurological disorders (FND).

In 2015 a book by the neurologist Suzanne O'Sullivan, a consultant at the National Hospital of Neurology and Neurosurgery in London, was published, titled *It's All In*

Your Head: True Stories of Imaginary Illness; an account of her twenty years of experience helping to treat conditions that exist in the murk between physical and psychological illness. One case, a woman called 'Mary', really stood out for me. Mary found her way to O'Sullivan because she would feel an overwhelming urge to close her eyes until, eventually, she was unable to open them at all. After numerous tests nothing was found to be physically wrong with her. That her husband was on remand for child abuse was not a relevant factor for Mary in what had been happening with her eyes. O'Sullivan treated Mary with muscle-relaxants, which seemed to take care of the problem. About a month later Mary was suffering with a profound amnesia and was readmitted to hospital. Doctors performed EEGs and brain scans and found nothing. However, a neighbour of Mary's mentioned to O'Sullivan that her husband had recently been released from prison. O'Sullivan ponders what Mary 'could not bear to look upon' or 'tolerate to remember'.[67]

In the past, Mary's presentation would have almost certainly been chalked up to hysteria. The same would have applied to a woman who had fits of uncontrollable shaking that seemed to come from nowhere and were unexplained by medical tests. The novelist and essayist Siri Hustvedt wrote a book about precisely this. *The Shaking Woman or A History of My Nerves* is a remarkable piece of work that is part personal investigation, part philosophical inquiry and a wider, searingly intelligent examination of how neurology and psychiatry have evolved over the last two hundred years. This book, along with Bessel Van Der Kolk's *The Body Keeps the Score*, deepened my wonder about the relationship between mind and body, feeding my ongoing inner question: can we actually separate the two?.

Hustvedt wrote her father's eulogy after he died. At his funeral, she read it 'in a strong voice, without tears'. In Minnesota, her home town, two and a half years later, Hustvedt was giving a talk on the college campus where her father had been a professor of Norwegian studies for forty years. Hustvedt gives a lot of talks and was feeling good about it. However, up on the stage, while talking about her father, she began shaking. Not just trembling but shaking, violently, from the neck down. Her knees were knocking, her arms were flapping. She could barely stay upright. Her mother was in the audience and described it as 'like an electrocution'. 'It appeared,' she writes, 'that some unknown force had suddenly taken over my body and decided I needed a good, sustained jolting.'[68] Then, every time she spoke in public, she started shaking again.

Was this a delayed grief reaction, or somehow connected to the migraines (and all the sensory upset that comes with them) that she had experienced since childhood? Becoming the determined detective in her own medical mystery, Hustvedt went in search of 'the shaking woman'. Was it epilepsy? Synaesthesia? Migraine? Anxiety? Grief? She had brain scans, endless tests, and entered into psychoanalysis – an enduring yet sceptical fascination of hers. Questions hatched from other questions. Hustvedt discovers that if she takes a beta-blocker tablet – medication used to treat high blood pressure and heart problems but also to help control symptoms of anxiety like shaking, trembling and blushing – before a public-speaking event she can almost avoid the shaking fits, but not completely. It distresses and galvanises her that she still doesn't know *why* the fits started.

Most fascinating to me is when Hustvedt begins asking where illness or pain begins and *she* begins. 'When I shook,' she writes, 'it didn't feel like me.' She explores how problems in the mind like depression, anxiety or any other mental health problem, as well as neurological illnesses, so often feel like invasion of the self. She wonders how much her personality affects her seizures and how much her seizures affect her personality. Instead of saying 'I am cancer,' she writes, we say 'I have cancer.' However, we do say 'I'm bipolar' or 'I'm epileptic'. It's true. I say, 'I'm anxious', 'I'm sad' or 'I'm tearful'. I don't say, 'I have anxiety', 'I have sadness' or 'I have tears'.

It is that 'I am' that I'm interested in when it comes to women's experiences of hormonal fluctuations, whether that's month-on-month with the menstrual cycle, during pregnancy, postpartum or menopausal. We cannot ignore biology, but there is something important to consider in the way we talk to ourselves and others about how we feel and how that might affect our sense of ourselves. The language I use when describing the way I feel with PMS is almost always definitive: I *am* sad. Not: I *feel* sad at the moment, recognising the feeling as a wave of emotion that will wash in and, eventually, wash out.

Does it make sense to ever describe some conditions as purely physical or psychological? As Hilary Mantel asked in her review of *The Shaking Woman* in the *Guardian*: 'How does physiology impact on personality? Where does the self begin and end? What is pain, and can it be abstracted from the body that suffers it, or the cultural context in which it is suffered?' Hustvedt is able to reach a point where she can conceptualise her seizures as part of her sense of self. 'I am,' she writes, 'the shaking woman.'[69]

I contacted Hustvedt through her agent to see if she'd mind talking to me for this book and, to my delight, she said yes. I called her at home in New York one afternoon. Her warmth and enthusiasm was immediate, as was her laugh. She laughs a lot, in big bursts, particularly after she's described something 'wacko'. We talk about how, as women, we can possibly begin to separate hormone activity from the self. 'When I immersed myself in hormone studies, what I discovered, number one, how much they don't know, but secondly how the research is just not modelled in enough of a social way,' she says. 'It's very important to understand that we're not just dealing with a car engine. We've been very successful in Western medicine in treating heart disease and that's because the heart is like a pump. The machine model works pretty well for heart function. I think it works for bones. After that everything else is just swimming because that reductive biomedical model of the body as a machine doesn't work. It doesn't work for the nervous system and it certainly doesn't work for hormones.'

I ask Hustvedt what her personal theory of functional neurological disorders – often referred to as psychiatry's 'blind spot' – is, and how it might relate to the wider picture of how women's distress is interpreted and treated. 'I think it's because of helplessness,' she says. 'There is a reason there are so many more women who have them in non-combat situations; people who are not fighting in wars or trapped in trenches, and it's because of helplessness.' In other words, what do soldiers after war have in common with women? 'Right. There was a famous book called *Shell Shock* written by a physician and psychologist called C. S. Meyers, who wrote the first paper on traumatised soldiers. He understood shell shock as a form of hysteria that had affected all these men and was very clear that the officers did not suffer from hysteria nearly as much as the soldiers. It's about a lack of power. Helplessness.'

Part Five

Part Five

Pain

In *Illness as a Metaphor*, Susan Sontag's influential text that challenges victim-blaming in the language often used to describe diseases and those who suffer from them, she writes: 'Illness is the night side of life, a more onerous citizenship. Everyone who is born holds dual citizenship, in the kingdom of the well and in the kingdom of the sick.'[70] Sontag wrote the book in 1978 while being treated for breast cancer. 'Although we all prefer to use the good passport, sooner or later each of us is obliged, at least for a spell, to identify ourselves as citizens of that other place.'

We all, at some point, know pain. It is a fact of life. It would be too simplistic to say that women hurt more than men; however, women are definitely taken less seriously than men when it comes to pain. There are studies showing that doctors, irrespective of gender, tend to under-treat female patients and take longer to administer medicine to them. A study published in the *Journal of Law, Medicine & Ethics* in 2003 titled 'The Girl Who Cried Pain: A Bias Against Women in the Treatment of Pain'[71] found that doctors often – incorrectly – believe that women have a 'natural capacity to endure pain', a product of the inherently high coping mechanisms needed for childbirth. Perhaps this is why the NHS routinely expects women to undergo procedures like hysteroscopies without offering any pain relief.

A hysteroscopy involves having a narrow telescope-like device passed through your vagina and cervix into the womb. The scope examines the inside of the womb and is used to investigate problems like unusual vaginal bleeding, heavy periods, pelvic pain, recurrent miscarriages, fibroids, polyps (non-cancerous growths in the womb). Treatment can usually be carried out at the same time, for example removing fibroids or polyps, adjusting or removing displaced IUDs (coils) or removing adhesions (bands of scar tissue) that might be contributing to erratic periods or fertility issues.

In the NHS guidelines it says that hysteroscopies are associated with varying pain levels: some women feel 'no or only mild pain', while for others it can be 'severe'. Although both local anaesthetic (injected into the cervix) and general anaesthetic can be offered by the doctors carrying out the procedure, there is no clinical requirement for this. The British Society for Gynaecological Endoscopy (BSGE), the organisation that oversees hysteroscopy, states in its guidelines that the procedure can be 'associated with significant pain'.[72] Yet many women seem to endure hysteroscopy without any anaesthesia whatsoever.

The procedure used to always be done under general anaesthetic, until daintier scopes entered the market in the early 2000s and women's cervixes no longer needed to be dilated. No general anaesthetic means you can be in and out of a clinic as an outpatient. The clear advantage of this is not having to stay in hospital. But until a woman's legs are in stirrups and the procedure begins, she has no idea how painful it will be. A friend of mine I'll call Holly described her recent experience of a hysteroscopy as 'breathtaking'. When she cried out in pain, 'fighting every urge to just pull the thing out of me', one of the nurses in the room asked her, with an air of languor, if she'd forgotten to take painkillers beforehand. At no point had anyone told her this would be advisable.

I looked at a number of NHS trusts' information leaflets and explanations of any potential pain ranged from being vaguely referred to as 'discomfort' to 'like period pain'. There was no uniformity in explaining pain-relief options. That the information could be so varied means that many women – placing their trust in healthcare professionals to tell them what might happen – won't have realistic expectations. Of course, you could argue that you don't want to scaremonger when it might not hurt at all, but women are saying that it does. In fact, a number of women have come together to form the Campaign Against Painful Hysteroscopy community and, with many patient testimonies available to read online, are putting pressure on medical authorities to take the pain women can experience with this invasive procedure seriously. In December 2018 the BSGE put out a statement in response:

> It is important that women are offered, from the outset, the choice of having the procedure performed as a day case procedure under general or regional anesthetic ... It is important that the procedure is stopped if a woman finds the outpatient experience too painful for it to be continued.

This might be encouraging to read, but what about all the women who have already been traumatised? Several women who have shared their experiences online speak of feeling the need to seek counselling. Elements of post-traumatic stress disorder (PTSD) are mentioned often. It is perfectly understandable that one would become nervous and untrusting in future medical settings. With such a lack of joined-up information for women, the assumption seems to be that if terrible pain is only ephemeral, it doesn't matter. Holly told me that, when she began to cry and shake, feeling as though she was in shock, the doctor performing the procedure and the nurse appeared unmoved. 'It was like they see women crying in that room all the time.'

NHS hospitals may not be offering women pain relief from the beginning for financial reasons. In 2013 the NHS almost doubled how much hospitals would be paid to carry out hysteroscopies as an outpatient procedure, during which women are usually not given pain relief. Cost-saving in a free national service is essential. Yet if women are saying they are being traumatised by pain, incentivising economy is happening at the expense of dignity. Hysteroscopy is an important procedure carried out all the time, and most women probably will find it tolerable. There is no way of knowing until it starts, however, and the whole experience would be better if we all knew in advance that we could ask to be sedated or receive pain medication. Even if it is inconvenient for a doctor to facilitate it, even if it takes more time, we *should* know that we can ask. It feels like a pretty basic right.

To paint a wider picture: a study from 2008 published in the *Official Journal of the Society for Academic Emergency Medicine* found that, compared to men, women often wait 16 minutes longer to get their pain medication in emergency departments and are 13 to 25 per cent less likely to be given opiate painkillers.[73] Could it be that their doctors have some inbuilt assumption that they're malingering? Or that clinical guidelines are overwhelmingly dependent on data from male bodies as the default?

Both. One of the major underlying problems in recognising female pain is the massive gender bias in medical research. The male body is the medical default. Clinical trials in medicine tend to include more men. Problems that only affect women, like period pain, are not given centrality. John Guillebaud, professor of reproductive health at University College London, says that some patients have described the cramping pain as 'almost as bad as having a heart attack'. Yet despite the vast number of women across the world suffering with severe period pain, the treatments are currently pretty limited. Painkillers, taking a contraceptive pill to reduce period heaviness or an IUD like the Mirena are about the size of it. Not for lack of scientists trying to offer something different, though.

With a team of researchers, Dr Richard Legro, who works at the Penn State College of Medicine, found that sildenafil (Viagra) can be used to treat period pain. In an interview with the website Quartz, he said, 'We published our results in a high-impact OB-GYN journal and we feel we made a major contribution to treatment that everyday practitioners could use.'[74] Much more research is needed before it could be approved as a treatment, however, in terms of dosage, means of absorption (oral or vaginal?) and longterm use. No one will fund this research. 'I've applied three or four times but it always gets rejected,' Legro says. 'I think the bottom line is that nobody thinks menstrual cramps is an important public health issue.' Clearly, without lobbies that champion the need for research into these aspects of women's health, less attention will be paid to the condition as something that affects women's day-to-day lives, and so the trickle-down effect of insouciance continues, to primary care doctors who just don't, or perhaps don't feel they should, take period pain seriously as a health issue.

A lack of female-animal models in research has led to a clear failure to scientifically study sex differences. Male and female immune systems are not the same, and yet the same medications are used to treat diseases because of this gender bias in research. In a paper published in September 2017,[75] Dr Susanne Wolf, a neuroscientist from the Max Delbruck Center (MDC) in Germany, showed that the underlying reason why male and female brains differ in their ability to fight the pathologies that cause nerve damage, such as Parkinson's disease and MS, can be found in the brain's immune cells. These are called microglia. Microglia function differently in men and women. Wolf's research found that men have more (and bigger) microglia, which probably therefore attack with greater strength. The male brain is not better at facing neurological diseases, but it is clear that men and women are equipped differently. The problem is that drugs haven't really been tested separately for men and women. Pharmaceutical companies privilege studies on male models. But if research only uses male models to research human disease and in the development of drugs to treat it, of course there is going to be a trickle-down effect on the way women are treated in medical settings.

I talked about the American pain medication study with Hustvedt. We shared incredulity at how women are 13 to 25 per cent less likely to be given opiate painkillers in an emergency department and she told me she had, in fact, been looking at the study only a few days previously. 'It's amazing that a man in pain simply doesn't have that stigma,' she says. 'A guy lying on the ground howling in pain is just somehow not a problem. Why would that be? I'm always asking myself that essential bottom question. What it is about?'

Believing

My own history of pain has, in many cases, been characterised by people not believing it. I began this book talking about the pain I've had with my periods since they began and how, from the very first one, I felt like the experience triggered a fundamental rewiring of what I knew pain was. As a seventeen-year-old, I experienced a kind of pain that went beyond anything I'd known.

My appendix became infected, went gangrenous and ruptured, causing peritonitis and sepsis. I very nearly died. I was in hospital for quite a while, missed massive chunks of school and acquired a medley of gut issues that have followed me into my thirties and, it seems, will never go away. This happened during a period of my early life that was pretty chaotic. To cut a complex and painful story short: there were separations, we moved around a lot and my siblings and I were quite unhappy. I desperately sought stability and affirmation, couldn't find it, and so, as a way of navigating the world, became completely self-sufficient. I stopped telling anyone how I felt. I left home at eighteen and never went back.

My 'presentation' of this intestinal explosion is foggy in my mind. Because I suffered so much with my periods and was having one at the time, my mum attributed my pain and need for horizontal-ness to that. This is a woman who would make us go to school unless we had a fever or were projectiling from mouth or bum. I wasn't taken to the doctor for this saga but she let me stay off school. It went on for three days: me lying on the sofa in pain with no appetite, sick bucket beside me for when the cramps really burned through my abdomen or back, groaning and sweating. I kept saying how much pain I was in and Mum, although caring, said, 'I'm sure it's okay.'

On the fourth day, I had a feeling between my legs that I'll never forget: a pain so violent it made me yell, then black out. Mum had gone to work for a couple of hours. I was frightened because my stomach appeared to be swelling quite rapidly. I don't remember her coming back but, when she did, she must not have liked the look of me because before I knew it we were in the car heading to the GP. We went in to see our usual family doctor, who I remember put his hand on my forehead and said, 'You feel a bit hot,' and told Mum it was maybe a virus along with a bad period. Thankfully she wasn't convinced and asked to see another doctor. This man laid me down on the bed, pulling a fresh sheet of that blue paper towelling over it, supporting me because I could barely move for pain at this point, and listened to my stomach through a stethoscope. Again, I don't remember this well, but Mum said he told her there were no bowel sounds at all and, coupled with my temperature, said it meant something bad was happening. I needed to go to accident and emergency immediately. He wrote us a letter that he put into a blue envelope, I remember that.

On the way to the GP surgery I had became incredibly, frighteningly thirsty and by the time we got to the hospital I had drunk about two litres of orange squash that Mum had made up for me. She had to carry me from the car to the front doors. The first thing the triage nurse did, after quickly wheeling me past a packed waiting room into a cubicle, was push the bottle away from my mouth. I had no idea why.

Turns out she had suspected sepsis, which is often character-ised by extreme thirst, and she was right. After what seemed like no time with the doctors, at which point the pain had moved into a metaphysical dimension – on top of the sharp baseline throb came a series of contractions that made me hang on to the bed, screaming like the girl in *The Exorcist* – I was in theatre. Hours later, many inches of bowel shorter and a couple of litres of pus lighter, I woke up on the ward covered in tubes big and small, hallucinating that everything was pink. I had no idea of the turmoil my innards and my mind would experience because of what had just happened.

Only in hindsight do I care to imagine what might have gone down if we had gone home after that first doctor had said it was just a bad period. I have a friend now who is a bowel surgeon and, a few years ago, I asked her. She said I'd have been dead on the sofa in a few hours.

Scars

As an adult woman, negotiating what my girlfriend calls 'poo chaos' has become almost quotidian. As a result of having a shortened bowel and all the rot that was swimming around in there, I have had a lot of issues with adhesions. These are fibrous bands that form between tissues and organs, often as a result of injury during abdominal surgery or from the very fact of having this type of surgery. Adhesions are thought of as internal scar tissue that connects tissues not normally con-nected. Over the years they have caused me considerable pain.

Between the ages of nineteen and twenty-five, I had several experiences of having to take myself to A&E with extreme abdominal pain, bloating, the inability to poo and, often, vomiting. Each time I would be X-rayed, deemed to have a 'partial blockage' of the bowel, admitted, made nil-by-mouth, put on a drip and told to wait for it to sort itself out. I wouldn't be allowed home until I'd 'opened my bowels'. The approach always worked after a few days. It was explained to me that, because adhesions can cause the bowel to twist and kink in funny ways, sometimes digestive matter can't pass through the bowel as it should, and builds up. Sometimes the bowel becomes completely obstructed and it's a clinical emergency in case it bursts and leaks into the abdominal cavity. (In which case you would, on some level quite satisfyingly, be completely full of shit.) I asked if they could operate on the adhesions to stop this pattern, but each time the doctors would say that research shows every time you surgically enter the abdominal cavity it can cause more adhesions.

After twenty-five these episodes started to get worse and worse. I'd try and wait them out, putting myself on 'bowel rest' – as advised by my surgeon friend – which meant having a liquid-only diet for a couple of days to try and shift things through my kinked bowel and avoid going to hospital. This often worked. I did not want to be lying in a strip-lit ward surrounded by nocturnal snoring and retching. The nature of the pain had shifted, though: as well as the blocked-up episodes, the searing cramps that often came after eating and the omnipresent nausea, I had chronic pelvic pain. My period pain seemed to get worse but it was this abstract, deep pain that got to me the most; it felt as though the bottom of my spine, my womb and bowel were being pincered together by the claw in a teddy picker machine. I went to my GP about it several times, who, upon looking over all my discharge notes, would say, 'I'm not sure what we can do at this stage.' Eventually, after one appointment where I ended up in tears, pleading for a referral, because the pelvic pain had kept me up most of the night, I was referred to a gynaecologist. When I saw him, months later, he told me he wanted to do a laparoscopy to have a look inside, and that many of my symptoms fitted with endometriosis. So they went in.

In the anaesthetic room I was so anxious my legs were shaking against the bed. I started crying, then apologising. The anaesthetist gave me something in my cannula 'for the anxiety' that felt, for a few seconds, like five thousand gin and tonics, and then I remember nothing. The gynaecologist didn't find endometriosis but did find that my small bowel, bladder, womb and ovaries were fused together with dense webs of adhesions. The back part of my womb was completely fused with my bowel. In my post-anaesthetic fog I remember the surgeon saying, 'Yes, it was quite a mess in there. No wonder you were in pain.'

The relief of those words.

I had 'proof' it hadn't all been in my head. This was the surgeon who observed that my fallopian tubes had been rendered all but useless by scar tissue and referred me for fertility treatment. If I hadn't insisted on being referred, I wouldn't have known I was infertile. How many years could I have spent trying to have a baby? How much additional pain had I been spared?

After this operation the obstructive episodes happened less and less. I didn't feel that deep pain as much for a couple of years. Then it started happening again. I was referred to a colorectal surgeon who said he wouldn't do any further surgery because of the risk of causing more adhesions. The pain continued. When I had my period the pain would rip through me, often to the point of vomiting. Sex could be extremely uncomfortable. The bloating and nausea after

189

eating came back. My baseline level of pain crept up. My emotional resilience for the pain dipped. Eventually, they said they would have another look inside and I was booked in for another laparoscopy. Again, they found that my small bowel and womb were fused with thick bands of scar tissue. This is similar to what happens to some women with endometriosis. I remember looking at pictures taken from the laparoscope and thinking they looked like squid. Lots of stretched, cross squid. I get confused about the timings, but I know that I felt considerable relief from pain for a while after this operation.

Two years ago, all the different manifestations of the pain crept back in. I found sex so painful sometimes that I'd started to become a bit phobic about it. My period pain would leave me flattened for at least one full day each month. I'd lie in bed at night trying to sleep, trying to shift my focus from the deep, sacrally-located ache as if to separate it from my consciousness. I didn't want it to be there. For months and months I did a good job of pretending it wasn't, taking pain-killers, altering my diet and generally trying to be as healthy as possible. It came and went, came and went, charring the edges of my sanity. My emotional response to the pain became more fierce. I would get angry with my body, my childhood, my mum, my GP, my friends for not seeming to take it seriously, my then-partner for not seeming to take it seriously. I sat in the bath with a belly like a woman at full term, folding myself like a pretzel with the cramps, feeling that familiar sense of grief for myself as a child that I seem to get when emotionally overwhelmed. I think of myself feeling important while helping to muck out the horses on the farm next to our house, running a flat-bristled brush over a tight, glinting haunch and the farmer's wife telling me I'd done a great job; finding woodlice and gently flicking them so they'd roll up into a little ball; being allowed to eat dinner in front of the TV; diving down in the local swimming pool deep end until my ears are raging. I imagine myself in the sun just having *no fucking idea*.

In October 2017 I had another laparoscopy. Even the thought of having to convince a GP for a referral exhausted me, so I paid to see a consultant privately. I'd heard about him being an expert in minimally invasive surgery and he said he had seen some patients experience relief with adhesiolysis (the removal of adhesions), but not many. Our conversation began with him asking if I really couldn't live with the pain. I said of course I could, I had been, but it was making life pretty miserable. He was willing to operate on the condition that I knew it *could* mean years of this pattern of surgeries, each time potentially causing more scar tissue. It was a risk I was prepared to take. Unfortunately it would have cost me £15,000 to do it privately, but he found a way of referring me back into the NHS. Once again, my organs were all stuck together. Again, I saw the pictures and saw strange marine creatures. I was in hospital for four days this time, because of the extent of the surgery, and the recovery was long. At the time of writing I am a year post surgery and, despite some interesting new scars, I feel okay. My periods are still excruciating but I guess it's a waiting game now to see what will happen.

What I do know is that I have sat in front of so many doctors now, almost all of them men, who have asked me variations on the same question: how much does it really hurt? I know that, after my egg-collection surgery at the end of my fertility treatment, I made four phone calls in forty-eight hours to the department, each time speaking to the same male doctor, saying I was in what felt like too much pain. I was lying on the bedroom floor in a ball with my back to the radiator and a hot water bottle pressed into my belly which was, by that point, swollen and tight like a drum skin. Each time I spoke to the doctor to say the pain was a lot more than 'period-cramp-like', he told me to take painkillers and ... use a hot water bottle. On the fourth phone call he sounded annoyed, giving me a 'most women have some pain after the procedure' speech. In the end I went to the walk-in centre at the hospital who transferred me over to majors. A gynaecologist did an ultrasound and discovered I had ovarian hyper-stimulation syndrome (OHSS). This is where the ovaries over-respond to the hormone injections used to stimulate the growth of the follicles that contain the eggs, causing enlarged ovaries with an excess number of follicles and fluid accumulation in the abdomen and, in rare cases, around the lungs and heart. On the ultrasound screen my ovaries looked like pitted hand grenades. I spent three days on the ward having to have my stomach measured every few hours by a nurse. Every wee had to be done in a measuring jug so they could check I wasn't dehydrated. I was fine in the end, and the nurses were – as they always are – funny, compassionate saints, but it just seemed like an avoidable experience had I been taken seriously in the first two phone calls.

I often wonder how different my medical history would look were I a man. It's impossible to know, but I have felt many times over the course of my life that as a woman I am expected to have this deep font of natural pain control; that I should be able to live with more pain than I can. I know that I feel butterflies every time I see a doctor about pain I'm experiencing, in case they don't believe what I'm saying.

Often I will underplay how I feel in order to sound more 'believable'. This is not wholly rational. Of course I have been believed and treated, in many cases, extremely well and with total respect. For some reason, though, when it comes to any pain between the ribs and pelvis, I fear I won't be heard. Obviously, there is a direct link between that first big dismissal aged seventeen (and the implications it could have had) and this fear. So much of my anxiety comes from 'what if?' questions relating to my bodily functions, consciousness and mortality. I know that this has been perpetuated over the years by conversations with doctors who have never met me before that moment, on that day, and make vast assumptions about what I should or should not be able to live with, when the only clues they have to my life are tiny, handwritten vignettes in a paper folder, from other doctors who have often only met me once. Many of whom have made me feel like they'd rather be anywhere else than sat across from me in a chair listening to me talk about pain.

While on the phone with Hustvedt I talked about some of this stuff and she said it had been the same for her and migraines. As a young woman, 'in the darkness of a year-long migraine', she says she saw a number of neurologists. 'I have to tell you, the contempt that I felt coming from them – they happened to all be men, but we can't rule out female contempt either – was like an ooze. It was like, "Oh, just a young woman with a headache? How disgusting."' She says it made her 'underplay' her symptoms, feeling a need to 'be tough and humorous as a way of fighting back, even though I was in a trough of despair. Obviously, this played into what I felt was almost as bitter, which was to be reduced to this snivelling female idiot.'

The recognition hit me like a bullet. So hard I burst out laughing.

Due scepticism

It is not a far-fetched claim or the bleating of a conspiracy theorist to say that modern medicine still doesn't take women seriously. We are a very long way from gender equality in research and treatment and the medical establishment deserves women's scepticism. Consider this: women can wait between four and ten years after first seeing a doctor with symptoms of endometriosis. The average is seven and a half years. In 2017, NICE released a new set of guidelines urging GPs not to overlook symptoms of the disorder, to try to speed up diagnosis.

Endometriosis is not rare. According to figures from Endometriosis UK, an organisation to inform and support women with the condition, one in ten women of reproductive age suffer from it worldwide, estimated to be approximately 176 million.[76] As a reminder, endometriosis is when tissue that behaves like the lining of the womb is found in other parts of the body. Each month these cells react in the same way as those in the womb, building up and then breaking down and bleeding. Unlike the cells in the womb that leave the body as a period, this blood has no way to escape, causing inflammation, pain and the formation of scar tissue. These cells can start covering the ovaries, fallopian tubes, bowel, bladder and stomach. It has even been known to reach the diaphragm and lungs.

Symptoms are terrible, including chronic pelvic pain, periods so painful women cannot go about their day-to-day activities, deep pain during or after sex and acute pain with bowel movements. Anxiety and depression, intense fatigue, feeling or being sick, irritable bowel syndrome-like symptoms (constipation, diarrhoea or a flitting between the two), bloating and back pain are also common. As the NICE guidelines say, many women report that the delay in diagnosis

194

'leads to increased personal suffering, prolonged ill health and a disease state that is more difficult to treat'.

No shit. Once diagnosed, there are several treatments that can help ease the suffering with endometriosis, such as pain medication, hormonal medications and surgery. As there is no definitive cure, though, helping women manage their symptoms must be an imperative.

As I described at the beginning of the book, hormone-like substances called prostaglandins are involved in period pain. Prostaglandins are made in every cell in the body and are released at the site of an injury to help with making a clot so the body can heal the injured tissue. They also cause blood vessels to contract, as well as muscle tissue, to stop blood loss. During our period (but also before), prostaglandins in the womb trigger muscle contractions as the lining gets ready to break down and then begins the process. Lots of prosta-glandins generally means lots of pain and, because they also stimulate the contracting of our gut muscles, can cause bowel changes. This is why we often get diarrhoea with period pain. In women who have endometriosis, all the extra prostaglan-dins being released where the rogue womb-like cells have grown can make other sites of pain in the body worse. If you have a disc injury, for example, it will most likely be much more painful around the time of the period. If endometri-osis is growing on the diaphragm, women may experience severe pain in the chest and shoulder area and down the arms. Endometriosis is also known to affect the sciatic nerve, which can cause burning pain in the buttocks and down the legs.

Being in that much pain for so long without being believed is going to do something to your constitution. Pain is exhausting. Having to explain the details of pain is exhausting. Having to continue living with said pain after you've asked someone to help you is demoralising. Perhaps we are just coming across as too stressy when describing pain, though, because research shows that, despite multiple trips to doctors and specialists, many women with autoimmune diseases or even multiple sclerosis can go undiagnosed for years because doctors tell them it's 'just stress'.

Even with diseases that affect men and women in similar numbers, women are often disadvantaged in their treatment. Take heart attacks, for example, which women survive less than men. Research done by the American Heart Association shows that only 39 per cent of women whose heart stops in a public place will get CPR. Men get it 45 per cent of the time and are a staggering 23 per cent more likely to survive.[77] Why is this? Perhaps bystanders are too scared to remove a woman's clothing? See too much hidden skin? Touch her breasts? Modesty eclipsing survival? Another problem is that women having a heart attack often present 'atypically', i.e. atypical for a man having one. Instead of the characteristic pressure in the chest, women may experience shortness of breath, dizziness, lightheadedness or fainting, pressure in the upper back or extreme tiredness.

Birth stories and pain thresholds

For some women, the experience of childbirth instils a deep distrust of medicine. Many mothers give birth in hospital settings in which they have little to no autonomy over their care and treatment. Given that childbirth will likely be one of the most painful and emotionally charged episodes of a woman's life, the potential for post-traumatic distress if the experience is unnecessarily traumatic and frightening is high – particularly if it doesn't accord with what the woman wants to happen. What looks like a 'normal' labour and delivery to a clinician can be experienced as deeply traumatic by the woman. The problem is that there is still much ideological stigma attached to childbirth. Many women approach it with fear, not having ever heard the truth of what awaits. As the journalist Eva Wiseman wrote in an *Observer Magazine* column on the subject of birth stories: 'Surely there are a thousand #MeToo moments happening in delivery rooms across the country, in part due to pregnant women's ignorance and fear. We don't know what's right, which means that, also, we don't know what's wrong.'[78]

Part of the problem has to be the squeamishness that surrounds birth stories. As Wiseman continues, there are similar absences in conversations across all aspects of women's healthcare: 'We are silent about our abortions, our miscarriages and our births, and sometimes it's because they were bad, sometimes because they were good, and sometimes it's because we don't think anyone wants to hear.' It is the female animal thing again. We keep so much of the blood, gore and fear that forms part of women's realities every second, every minute and every day across the world quiet in case it upsets or disgusts other people. A glance at the conversation threads on sites like Mumsnet tells you that women are dying to tell their stories because the experience is just too much to keep in. They want to see their pain, trauma, fear and delight reflected back at them.

There is much ideological debate surrounding childbirth today, as there always has been, much of it bound by sexual politics. Giving birth is a moment in a woman's life where she is confronted by her own biology in Technicolor. The hospital environment is a mirror to her division by biology: most midwives are women, but most doctors in senior positions are men. In some cases this is a power dynamic ripe for creating problems. You only have to look at Mumsnet for five minutes to read about doctors not listening to women at what is undoubtedly going to be the most painful, scary and utterly bewildering experience of their lives, and the knock-on effect it had. Women make different decisions about giving birth based on personal, cultural, social, medical and many other factors. Some women elect to give birth by caesarean section (CS), as has been stipulated as a right within the NICE guidelines for eight years. There is still stigma around this. Since the phrase 'too posh to push' was coined, about twenty years ago, as a slur against women (usually those who were well-educated or rich) who simply cannot face the chore of pushing a baby out through their vagina, it has stuck. Some women find giving birth much harder than others but there is no 'easy' way to do it and the evidence is complicated about what is safest. Sometimes what is 'better' for the mother isn't better for the baby and vice versa. But in 2014, with an increased number of caesareans, infant and neonatal mortality was at an all-time low.

Journalist and campaigner Rebecca Schiller, author of *Your No Guilt Pregnancy Plan,* previous director of the Birthrights charity and a doula, wrote in a piece for the *Guardian* that 'the desire to promote a birth ideology tied to something that worked for us personally – or the attempt to steer others away from traumatic experiences we have had ourselves – is seductive and understandable'.[79] But in being singular with ideas we run the risk of stopping progress on the most important thing: women's complex, individual needs being heard. Many of the women Schiller helped with Birthrights were hoping to give birth by CS because they had endured 'previously traumatic vaginal births, physical or mental ill health, or are survivors of sexual abuse'. Others had examined the (far from clear-cut) evidence available and made an informed decision that a CS would give them, and their baby, the 'best chance of an emotionally and physically healthy start'.

Part of the problem, too, is the way we fetishise the 'natural'. I was at a dinner party once with a woman who said that she believed every woman should be made to give birth naked in a forest, with no painkillers, and 'get back' to the 'natural' way of our 'foremothers'. I've never forgotten it. Hard to know where to begin, really. First, all women have different pain thresholds – one of those things based on the complex constellation of genes and life experience. Some women find having a coil fitted distressing (hi!), others don't. Some women have period pain (again: hi), others don't. What *makes* some women experience different types of pain to others is largely a mystery, but should we really care? Sometimes we cannot help responding to someone else's distress with a version of our own, and it's not always helpful. Recently, an old friend made contact again after some years and we arranged to meet up. On the day, I was catatonic with period pain. I said I couldn't make it, suggesting other dates. Her reply went something like: 'That's a shame. I've never had period pain!' I didn't know what to do with that information. Offer congratulations? Condolences? Just because we can't relate to how someone feels doesn't mean they don't feel it, yet we almost wear our version of someone's pain, whether we mean to or not, as a badge of honour.

Back to this woman's mid-tagine theses. Should we assume there is a reason why she doesn't take into account how many mothers and babies died in childbirth in the past without medical intervention? Or how many continue to die in less developed countries? World Health Organisation figures on maternal mortality state that almost all maternal deaths (99 per cent) occur in developing countries.[80] More than half of these deaths occur in sub–Saharan Africa and almost one third occur in South Asia.

I continue to be fascinated with our preoccupation with what is or isn't 'natural'. I have heard well-educated, right-on feminist female friends say they'd 'rather' do it without any medication. All power to them, of course, but why? What informs that desire to do this huge, painful, life-altering thing without evidence-based interventions that might make your suffering less?

We still attach so much nobility to suffering, particularly that which women experience. Where does it come from? The idea that we've suffered so much throughout history that to be seen to be vulnerable or reliant on anything or anyone but our own grit, that deep font of natural resilience to pain, is somehow 'less than'? Is knowing, feeling, screaming, enduring physical pain, even if we don't have to, the marker of a stronger, more powerful woman? What if we could – as charities like Birthrights encourage – move towards a more unified, compassionate understanding that the most powerful thing a woman can do is be informed, know herself, know her limits, know what she needs, ask for it and be respected for asking. Wouldn't that help all of us?

A word on wellness

The 'holistic' world of reiki, essential oils, herbs, meridian lines and crystals is nothing new. It has always existed, invariably used by those who have the money to not solely rely on state healthcare. But in the last few years there has been a more powerful shift in the way women conceptualise and treat their health. According to figures from the Global

Wellness Institute, the global wellness industry grew 10.6 per cent from 2013 to 2015: from $3.36 trillion to a $3.72 trillion market.[81] More people than ever are 'detoxing' (a complete fallacy: our liver and kidneys are detoxing what we consume all the time anyway), plugging in salt lamps, going to oxygen bars and eschewing 'traditional' medicine.

Actor Gwyneth Paltrow's lifestyle brand Goop has undoubtedly been a major player. In its newsletters and on its website, reaching millions of women, Goop has evangelised 'treatments' ranging from crystal therapy to vaginal steaming as a way of optimising the health of your reproductive passages, to $66 jade eggs to be 'worn' inside the vagina – purported to help with anything from hormone levels to bladder control. Very little of these treatments have any basis in evidence that isn't anecdotal. But while it's easy to scoff at Paltrow, Goop or anyone else in the wellness industry making dubious, cure-all claims, and definitely wise to be sceptical about the risks, it is a bit more sophisticated (and less snobby: the placebo effect can come from all sorts, if you can afford it) to consider what are the driving forces behind the rise in consumption of these theories and products – invariably among women.

The ascent of the wellness industry is a clear response to a medical establishment that so often dehumanises and dismisses women. Tired of being dismissed, unheard, left in pain or forgotten, tired of being tired, women in the Western world have created their own alternative healthcare system. The wellness industry has the female body at its centre, providing soft-lit, embracing and safe spaces for women that make them feel seen, heard and, crucially, an individual with individual needs. As the journalist Annaliese Griffin writes in a piece on wellness for Quartz: 'It's easy to see how a sympathetic reiki practitioner who listens closely to your descriptions of what it feels like to inhabit your body, and vows to help you feel better, could seem appealing.'[82]

Whether it's a biodegradable cup of cold-pressed beetroot and ginger juice, light therapy, sound baths, coffee enemas, vitamin drips, chi, shamans, meditation, acupuncture, yoga, yoni crystals or leech therapy, wellness services and products are designed to make women feel unique and treated as such. We know that meditation, yoga, acupuncture and massage all have an evidence base – particularly meditation, which research now shows can actually change the physical structure of emotion-processing areas of the brain when practised regularly. Also, there are many practitioners in this field who have been committed to education and are kind, dedicated people potentially making their clients feel a lot better about themselves. As many have pointed out, though, much of the wellness industry hitches a ride on what Griffin refers to in her piece as the 'coattails of compassion and competency, benefiting from the utter lack of warmth found in mainstream medical treatment'.

One of Paltrow's most vocal critics is the San Francisco-based gynaecologist and obstetrician Dr Jennifer Gunter. Gunter started blogging in 2010, focusing on trends and products that are marketed to women to improve their vaginal health, improve their sex lives or help them get out of pain. 'Every single day I am talking to women about how you shouldn't use this product and why you shouldn't use that product,' she told the *New York Times*. Along with running a busy clinic, Gunter routinely responds to Goop's decrees on women's health. She has written some posts that have gone viral that Paltrow et al. largely seemed to ignore, until she posted a meticulous, data-heavy rebuke of Goop's promotion of a diet low in lectins – the carbohydrate-binding proteins found in beans, legumes and some grains – for women. Goop came back with their own post, which Paltrow tweeted to her millions of followers, saying, 'When they go low, we go high.' (A line borrowed from a speech Michelle Obama made.) The post referenced Gunter in all but name, referencing the 'wielding the lasso of truth' line she had used. They suggested she was continuing to 'critique Goop to leverage that interest and bring attention' to herself. It continued by referring to Gunter's 'strangely confident assertion that putting a crystal in your vagina for pelvic-floor-strengthening exercises would put you in danger of getting toxic shock syndrome'. She came back with her own post, 'Goop's Misogynistic, Mansplaining Hit Job', writing: 'I am not strangely confident about vaginal health; I am appropriately confident because I am the expert.' She cites her qualifications: a medical degree from the University of Manitoba, a residency at the University of Western Ontario and a fellowship in infectious disease at the University of Kansas Medical Center. Were I on the team at Goop, I'd be pretty wary of arguing with her.

'I don't want to put hormones in my body'

If we start believing that science and doctors are not to be trusted, wherever your lack of trust comes from, we easily open ourselves up as a honey trap for misinformation. It might be starting to obsess about parabens in shampoo, the inflammatory properties of gluten or dairy (when we're not coeliac or lactose intolerant), 'toxins' in everything from suncream to cough mixture, or even declining to vaccinate your child. I have lost count of the number of times I have heard women around me say 'I don't want to put hormones in my body if I can help it' with regard to contraception, trying to manage heavy periods, or PMS. It's easy to see where the fear comes from.

There are many known side-effects with hormonal treatments (weight gain, skin problems, mood disturbance, gut issues) and we also know that, because options are limited when it comes to women's health, doctors often put a woman on medication in a leap of faith, hoping that it might help. What's important to remember is that, often, it does. The contraceptive pill is recommended by NICE as a first-line treatment for heavy periods and PMS because it continues to have a good evidence base. Some women tolerate it better than others. Unfortunately, due to the huge differences in physiology from one woman to the next, we just can't tell whether it will work for us until we try it. Nor do we exactly know why some women react so badly. But it's the language around this fear that really interests me.

We reject the notion of putting 'hormones', the word acquiring a certain dirtiness, into our bodies; as if hormones aren't one of the most important parts of what keeps us alive. Perhaps it's the idea of foreign agents; synthetic versions of what we naturally have that make women fat, mad and sad all the time. Again, it is an understandable fear, but there is a connecting thread, I think, in this growing fetishisation of 'natural'; keeping our bodies as 'clean' and pure as possible, because that's what has been marketed to us as an ideal. 'Clean' is a particularly pernicious term because, while regulatory bodies have started to watch companies using 'natural' as a marketing tool much more closely, 'clean' is far more nebulous, less open to challenge. All these terms rely on tapping into or creating *values*. It gets more sinister the more you think about it, particularly when you consider who is ultimately profiting in the industry – white, rich people. Particularly if those people purport to care about women's health yet don't stand up to the systems whose policies continue to oppress women. In reference to Paltrow not publicly opposing the US administration, Dr Gunter tweeted: 'You know who profits from ignorance and the patriarchy? Gwyneth Paltrow.' 'Pseudoscience is the patriarchy's handmaiden,' she wrote in another. This was such a striking phrase that I got in touch with Gunter by email and asked her to elaborate.

'Pseudoscience serves the patriarchy by misinforming women,' she told me in her reply, while working on the edits for her own book, *The Vagina Bible*, out this year.

> The less knowledge women have about their bodies and how to treat illnesses, especially the less they know about their reproductive tract, the better it is for a patriarchal society.

How so?

> As women typically make less than men, even small amounts of money spent on useless products (never mind the impact of the misinformation) will carry a greater financial burden. Women who are scared into the idea of 'toxins' in tampons either spend more money cumulatively over their lifespan on 'natural' products ($2–$3 a month over 30 years can add up!) or are afraid to use tampons altogether.

Gunter also wanted to stress that 'reproductive myths and lies about contraception' could increase the risk of an unplanned pregnancy.

Lies about IUDs being abortifacients [something that will cause a miscarriage] are common and there is even one family planning app that overstates its efficacy [Daysy – as I explained earlier].

The 'burden of unplanned pregnancies', she says, 'falls on women'.

We must not forget, too, that even if something is helping someone without causing any harm, most of what the wellness industry has to offer is only for people who have money. Access to healthcare across the world depends on socioeconomic status and race. Receiving kind, considered healthcare should not be the privilege of those with private health insurance or money to burn on colonic irrigation. There have been campaigns in both the UK and the US to promote compassion in clinicians, such as the 'Hello, My Name Is' campaign started by the late Dr Kate Granger while she was a terminally ill cancer patient. During a stay in hospital in 2013 with post-operative sepsis, Granger observed that many of the staff caring for her did not introduce themselves. It felt incredibly wrong to her that such a basic step in communication was missing. Using social media in the first instance, Granger and her husband Chris Pointon began campaigning to remind healthcare staff of the importance of introductions and considered communication. Granger was awarded an MBE for her work and, since she died in 2016, Pointon has continued speaking at conferences across the world to keep the campaign alive. I followed Granger on Twitter and read her updates with interest. As someone who has spent quite a bit of time in hospitals over the last few years, I have noticed a real change in how doctors and nurses approach patients before beginning treatment. Nine times out of ten you will learn someone's name before they push a thin tube of metal into one of your veins, or squirt lubricant onto a speculum and slide it into your vagina under a spotlight. It does help.

Healthcare professionals would do well, it seems, to relearn some of the basics when it comes to interacting with another person. Part of the reason people are turning towards leeches and flotation tanks is because doctors have not adjusted their posture to make eye contact, shaken hands or listened attentively to someone's story. It is not just a sensible financial strategy – treat someone well and thoroughly in the first instance and, hopefully, they'll have less need to come back or complain – it's the human thing to do.

How to treat a problem that is neither 'just hormones' nor 'all in the mind'

Before I began writing this book one of the first things I wanted to look into was whether there could be an evolutionary function to PMS. It's something my female friends and I talk about a lot: if we can feel that sad, seemingly from nowhere, there *must* be a reason for it. What in our genetic make-up, our evolutionary history, could benefit from the sadness, anxiety, anger or reassurance-seeking we feel with our hormonal fluctuations, or the uncomfortable buffet of physical symptoms? It's an interesting question to which the answer is, of course: we don't really know.

Research has thrown up some ideas, mostly focusing on whether PMS could have a selective advantage by increasing our chances of pairing with fertile partners. One literature review study from 2015 by Professor Michael R. Gillings of Macquarie University in Sydney, Australia, published in the journal *Evolutionary Applications*, suggests that PMS may serve to improve the reproductive outcomes of women. 'Animosity exhibited during PMS is preferentially directed at current partners; and behaviours exhibited during PMS may increase the chance of finding a new partner,'[83] Gillings writes. In layman's terms, we might bite our partner's heads off if they haven't impregnated us yet because of a primal, subconscious push to piss them off so much they leave us, giving us more opportunity to find someone who will.

Gillings's study also explores the gene heritability of PMS and suggests that understanding this 'evolutionary mismatch' might help 'depathologise' PMS. To me, the notion that my PMS is rooted in my absolute, fundamental role of having babies feels a bit heartbreaking. Of course, as with any matter involving hormones, it would be remiss to dismiss biology altogether. But the study suggests that women may find a 'solution' to their 'dilemma' in cycle-stopping hormonal contraception because it would help mimic our 'ancestral state'. That is: women in hunter-gatherer societies were likely to be pregnant, or if caring for a baby, in a state of 'lactational amenorrhoea' (the temporary postnatal infertility that occurs when a woman is breastfeeding and isn't menstruating) for most of their reproductive lives. Consequently, they would experience fewer menstrual cycles.

While cycle-suppression may be an understandably attractive idea – and, indeed, effective treatment – for women with PMDD, it also makes me think about Jane Ussher's work in questioning whether, in giving us theories that involve our genes and wiring, then offering us a simple pill to make it better, this approach is pathologisng our distress. Saying it's something we can't really help because it's what we've been programmed to do implies a lack of control. There's something to be said, though, for women seeking clear explanations for their distress. By looking for evolutionary function, we're looking for something to hold on to, because feeling sad, anxious or angry is not comfortable; and, as we're women with a history of having our emotional selves oppressed and contained, we fight and question the way we feel because, on a subconscious level, we feel like we shouldn't be feeling it.

When I first went to my GP to talk about managing my PMS, I suppose I was looking for a panacea. I just couldn't reckon with the idea that I might slip into that way of feeling every month until the menopause. That's a lot of days. Every month I would forget and then: *woompf.* Then I'd think, I have to do something about this. Once I started tracking my cycle with the Clue app, being able to predict when I might feel crap, or, looking at it when I *am* feeling crap, being able to see that I'm ovulating or entering the PMS phase, was a big help. It took away some of the 'from nowhere' feelings and made me feel more like a body of rhythms. I have tried some other strategies, too, but never found my panacea because no such thing exists.

Pills, gels and coils

Initially my GP prescribed Cerazette, the 'mini-pill' that only contains progestogen. Her rationale was, given that my symptoms are worst around ovulation, stopping ovulation altogether (as the mini-pill can do) 'might' work. I am more aware than ever of the weight of that 'might' that comes with anything related to the ways of the mind, but I wasn't entirely convinced. I took a few days' worth because my partner said I might as well try and I agreed, but after doing a literature search and finding meta-analyses that do not support the claimed efficacy of progesterone in the management of PMS, I stopped. My GP also suggested that vitamin B6 might help. I'd heard this before and bought some from Holland & Barrett, sticking with it for about three months. I felt no discernible difference whatsoever. As may be expected, any conclusions about B6 are limited by the low quality of trials. It's the same story for supplements like evening primrose oil, St John's wort and agnus castus. Some small studies show some effects, but taken overall trials show conflicting results. The evidence is not good enough. That's not to say some women may not feel an effect – there's a reason these products end up being sold. Placebo may be a big part of it, but if what you're taking is not causing any harm, there's nothing wrong with that.

I'd read a lot about a man called Professor John Studd, an eminent consultant gynaecologist who has been publishing research on PMS for twenty-five years. Mr Panay trained under him. Studd treats women in his clinic with oestrogen gel and supplemental progesterone. The continuous oestrogen suppresses the cycle and the progesterone acts as a protector for the womb lining. Women apply the gel to their thighs every day – the same oestrogen gel that is used in HRT – or have the option of an implant. Progesterone tablets are taken for around one week in each cycle. Curious, I rang Studd to talk about it when I was writing a piece about PMS for The Pool. He has the kind of booming, posh voice of a cartoon headmaster. Telling me about the 'eight to ten women a day' he sees in his London clinic, he says they've 'been through everything – antidepressants, mood stabilisers, psychiatrists, you name it, because psychiatrists are scared of hormones' and that the treatment he provides has 'very few failures'. But what about the theory that women who have bad PMS are sensitive to their own levels of oestrogen and progesterone?

Won't introducing more make them feel worse? 'Severe PMS is due to intolerance of progesterone in the last half of the cycle. In a 28-day cycle, the worst days are day 14 and 21. It's all so obvious. But if you're sensitive to progesterone, some PMS symptoms will come back when you take the tablets,' Studd says. For women who still find these intolerable, he says a Mirena coil may be a better option than tablets, as the progesterone is metabolised differently by the body when it is released directly into the womb.

I was convinced, but couldn't afford Studd's high fees. I'd looked on Mumsnet and seen many desperate women willing to travel across the country and pay in excess of £600 to see him. However, Studd does not 'own' these treatments. They *are* approved for use within the NHS – accessing them is the problem. Not a first-line treatment in current NICE guidelines, they'll likely only be prescribed when you are referred to a gynaecologist. My PMS did seem to be getting worse as I got older and I was open to trying anything, so I went to see my GP and told her about this approach. She'd never heard of it, but cheerily offered to look into it for me. After conferring with a gynaecologist at the local hospital she called me back to say we should try it.

I booked myself in to have a Mirena fitted, scared of the idea of taking tablets that might induce my worst PMS symptoms, also hoping that it might help with my heavy bleeding as well. The experience was not awful, but not breezy. In order to fit the coil, the doctor has to clamp your cervix, inject it with anaesthetic and then insert a plastic and metal device into it. The whole thing must have taken less than fifteen minutes and the doctor, a soft-voiced Australian man called Brendan, could not have been more gentle or respectful, talking me through each part of the procedure

211

before he did it. As he inserted the Mirena, the chorus of Carly Rae Jepsen's 'Call Me Maybe' kicked in on a radio in the background somewhere and I suddenly felt very sick. My ears rang, my head swam. I told Brendan. 'Ah, yeah, that sick-feeling isn't something in your control,' he said. 'Take as long as you need.' The nurse passed me one of those NHS cardboard vomit-catchers that look like trilby hats. Luckily I didn't need it. As I started to feel better, Brendan explained that the cervix is linked to the vagus nerves, which control heartbeat, breathing and blood pressure. When they're over-stimulated in women with a sensitive cervix (I have one, evidently) they can cause a sudden drop in blood pressure and make you feel faint. I went home afterwards having cramps, which Brendan told me to expect, but they basically didn't go away. I started applying the oestrogen gel to my inner thighs every day, incredulous at the idea that a couple of drops of cold goo on my legs would do something.

I stuck with this dual treatment for six months, but thought about stopping many times. The cramps from the Mirena were intermittent but persisted. I bled, thinly but noticeably, constantly. I was bloated, my breasts went up a cup size (great, you'd think, but when they're sore to the touch no one really gets to enjoy it), I had headaches and felt nauseous *a lot*. All these are documented side-effects from the oestrogen gel and Mirena. I went back to the genitourinary medicine (GUM) clinic a couple of times to ask about the bleeding as it just didn't stop, and each time they told me to stick with it. I knew that the Mirena could cause heavier or more irregular bleeding for the first three to six months, but I didn't know it would be constant. I also really, really *wanted* it to work.

For a little while I thought I'd noticed a difference in my anxiety levels, but now that the gel and coil had taken over my cycle in ways that seemed mysterious and unpredictable, although I knew that I had to allow for a 'settling-in' period, I found myself missing being able to look at my little Clue chart and know why I might feel a certain way at a certain time. This felt too unpredictable. Oceanic. I couldn't keep going. I was tired of feeling sick and dealing with the constant blood loss was becoming very boring. Also, I think any perceived change in my anxiety levels may have been a product of blind faith; a deep will for something to work or take away this monthly slide into emotional entropy.

I kept googling 'Mirena anxiety' or 'Can Mirena make PMS worse?' and the results were a mixed bag. Lots of women swear by it, saying that after the settling-in period (up to six months of your life, no big deal) they felt better than ever. I tried to hang on to this. Yet lots of women also said that the Mirena made them feel more anxious and 'moody' than before, the theory being that, if you are sensitive to your own fluctuating levels of progesterone, even a small device releasing low levels of the hormone only into the womb might affect you. But I'd been told by my GP and the GUM doctors that any mood-related effects were 'highly unlikely' because the Mirena's hormones are only locally released. Hm.

One morning, six months in and what seemed like hundreds of pounds of sanitary-wear-related spending later, I thought: enough. I did not feel better, I felt worse. All these new, open-ended physical symptoms were grinding me down. I threw my bottle of gel away, went to the walk-in GUM clinic and asked for the Mirena to be removed. After a short 'Are you sure you don't want to persevere?' conversation, a nice man slid a speculum into me, grabbed a thin, barbecue-tongs-like device and quickly tugged the coil from my womb.

It took about thirty seconds, feeling utterly alien but not painful. I asked to see the thing. In a little cardboard tray sat the T-shaped device; gleaming white and dotted with blood, looking like something from a box of fishing tackle. I asked the doctor what he thought about whether the Mirena could have made my anxiety worse and his reply was very different to the other doctors'. He said that, even though the Mirena only releases small amounts of synthetic progesterone into the womb, small amounts are absorbed by the rest of the body and can cause side-effects, particularly in women who are sensitive to their own rising levels of progesterone each month. 'Huh,' I said. 'I wish someone had told me before.' All those 'stick with it' conversations. The doctor was sympathetic, making a joke about invoicing them for all the sanitary towels, and said that the medical profession often expects women to 'put up with' a lot. It was heartening to hear him say that.

I left the clinic relieved, but also feeling a strange sense of having been meddled with and slightly afraid that I had run out of options. I knew I could try the GnRH analogue drugs that would completely suppress my cycle and create a synthetic menopause, but, with their considerable side-effect profile (menopause-like hot flushes, headaches, low libido and vaginal dryness), did I really want to? What else could I do?

It occurred to me that I had never really discussed the patterns of my cycle in any depth with a therapist. At the time I didn't have one. I put off the idea, feeling more at sea than I had when I started. When I went back to see my GP to discuss what to do next, she did suggest the GnRH analogues. 'I'm not sure what else there is at this point otherwise,' she said. 'You already take an SSRI, which would be another option, but . . . ' She did a sort of half-smile. A kick in my gut just said: no. But where did that leave me?

Tomatoes, microbiome and yeast

I have always eaten well. Cooking is my great joy and I see most fruits and vegetables as deities. Through process of elimination over the years I have learned what sends my sensitive gut into bloated, indecorous meltdown: all beans except black, onions (the worst), too many wheat-y goods, barley, red meat, too much chilli, deep-fried things and, to my utter heart-break, raw tomatoes. I have no clinical intolerances but all these things appear in IBS-irritating food lists and, by virtue of having an enormous appetite for both eating and cooking, I have observed what does and doesn't 'suit' me, shall we say.

Other than treating beans and onions as biohazardous – it's better for everyone – I eat all the other things in moderation. I would not stop eating raw tomatoes if you paid my rent for the rest of my life, though; there is little finer in this world than a perfectly ripe, sun-warmed tomato clicked off its vine, cut in half, anointed with a few grains of sea salt and slid into your cheek. If the tomato is right, the bursting of its sweet-yet-utterly-savoury membranes against your teeth is almost pornographic.

Anyway, I'd read a lot about how the notorious 'diet and exercise' combination can help with PMS, but as someone who knows food, largely knows what's good and not-so-good, and cooks every day, I thought I had the diet bit sussed. I have wilfully ignored anything or anyone extolling the benefits of a sugar-free diet – when it comes to women's health, there is a lot of this about – or, indeed, any anything-free diet unless you have a clinical intolerance. The layers of puritanism, smugness, misinformation and privilege within the clean-eating world make me shudder. But I wondered whether I was missing something in relation to my PMS. Seeing as I am, after all, an organic body kept going by organic matter, the nature of that matter will have an effect on how I work.

Given my longstanding interest in the gut, I have been reading about microbiome for a couple of years now. Our bodies contain a vast array of micro-organisms. We are full of bacteria (each gut contains about 100 trillion) but also carry around millions of single-celled organisms known as archaea (microbes with no nucleus), as well as viruses, fungi and other microbes. All together, these are known as the human microbiota. 'Microbiome' stands for all the genes contained in our microbiota and is as unique to an individual as a finger-print. Research has shown that they have so many important functions within the body that the microbiome is now being referred to as a 'virtual organ' or 'the second brain'.

Along with assisting with digestion and synthesising the nutrients we take in, our microbiota release chemical profiles that regulate our immune function, metabolism and, most interestingly to me, mood. They also help to control the level of oestrogen in the body. The subset of the microbiome involved in metabolising oestrogen has a wonderful name: *estrobolome*. Lovely to say out loud. Science is focusing more and more now on isolating the mechanisms by which our microbiome is connected to our overall health. Imbalances in the microbiome have been linked to digestive problems like IBS, to fertility, immunity disorders, obesity, Alzheimer's, arthritis and, more prosaically, energy levels. I kept seeing a man called Peter Cox pop up in articles on microbiome here in the UK. Cox is a physiologist turned clinical nutritionist with a significant interest in the relationship between our microbiome and mood. I'd read that he can analyse your urine or stools to get a measure of your microbiome and how your health may be affected or improved with dietary changes. On his website, he says he has helped many women with premenstrual syndrome feel better with nutritional changes. I made an appointment to see him in the room he rents from Biolab, a nutritional biochemistry medical laboratory just off Tottenham Court Road, London.

Cox is a gentle man, so softly spoken you often can barely hear what he's saying. Nevertheless, his detailed questions and smiliness made me feel like he was genuinely interested in helping me. After I'd reeled off my list of symptoms – the premenstrual anxiety and tears, pain, constipation, bloating, etc. – he offered a working theory that, along with potential dysbiosis (an unhealthy microbiome) and excess yeasts, I had impaired oestrogen metabolism. He'd explain more after I'd done something called an Organix Comprehensive Profile nutritional test to look at my cellular metabolic processes. I thought this would be via a stool sample, mentally preparing myself for a wincing toilet experience, but our metabolic efficiency can be analysed through our other waste product: urine.

Obviously I'd never heard of this test. As far as I could tell from my online research, it is a sophisticated urine test mostly used by elite athletes to determine exactly what they should be eating to optimise their physical performance. It would cost me an eye-watering £279 (on top of Cox's £150 first-appointment fee) and be processed by a US lab called Geneva Diagnostics. Cox said I would receive all the material in the post and should pay the lab directly. Sure enough, a package arrived in a few days containing a specimen pot, a silver 'biological specimen' bag, an ice pack and a marked envelope to send back to the lab. I was to produce a 'mid-morning, mid-stream' sample before eating or drinking. Easy enough. Off it went, along with a slip for my credit card details. After two weeks Cox had my results. On the phone he said, 'This is very interesting, Eleanor!' I heard him taking a gulp of water. I shifted in my chair.

Back in the room at Biolab, full of anatomical models, Cox greeted me with an extremely long handshake before reading me my findings. 'Your Organix test results show a series of abnormalities which help describe the relationship between your gut symptoms and mood,' he said. 'Most importantly, you have an overgrowth of lactic acid producing bacteria in the gut, which is associated with increased gas and bloating, especially with fibre-rich foods, but also the production and release of toxic substances which are likely to contribute to symptoms of poor digestion, including an offensive odour, abdominal pain and nausea.'

I quietly pondered the last time I'd had a takeaway curry.

'You also have an excess of something called arabinose, which describes an excess of yeasts which ferment sugar, causing you to produce excessive gas and bloating with the consumption of sugar and carbohydrate-rich foods.'

Yikes.

'The yeasts will also synthesise ethanol which is metabolised to acetaldehyde. These are the substances associated with feeling drunk and hungover, and it is likely your mood reflects these toxic effects. It is also likely these substances impair your cognitive function, and you are likely to feel tired and moody with the consumption of excess carbohydrates.'

It was a lot to take in, but he had more to say yet.

'It would need a stool test to confirm, but it's likely these results show a dysbiosis, yeast overgrowth and SIBO (small intestinal bacterial overgrowth). These are not only associated with unpleasant gut symptoms but also distorted signalling to the brain, associated with fatigue and mood issues.'

'Right,' I said, holding my middle. And what about my hormones? What can my yeast overgrowth tell me about PMS?

'Okay, well the yeast overgrowth also triggers cravings for the sugars which feed the population of yeasts, found in carbohydrates and sugary foods. This disrupts blood sugar levels, causing over-production of insulin and hormonal imbalances, including the over-release of the stress hormone cortisol and the increased uptake of oestrogens.'

I asked him to break down the oestrogen aspect further. He explained that after oestrogen is made in the ovaries, it circulates through the body to the womb, breasts and other organs, until it reaches the liver. Here, it is deactivated or 'detoxified'. Deactivated oestrogen is sent to the gut, where it should stay deactivated so it can leave our body in our poo.

'This is healthy oestrogen metabolism,' says Cox. However, when dysbiosis is present, it goes a bit differently. Unfriendly bacteria make enzymes that reactivate oestrogen in the gut. This is problematic because the reactivated oestrogen re-enters the body and causes excess oestrogen levels. 'This is impaired oestrogen metabolism.'

So, in essence, a badly behaving microbiome can cause more oestrogen to stick around, which can have an impact on the way our brain works given what we know about hormones interacting with our neurochemistry?

'That is an accurate summary, yes.'

And if I am constipated a lot, in theory I am 'hanging on to' too much oestrogen?

'Yes.'

So what should I do?

Cox said that, although the findings and explanations sound complicated and daunting, the remedies are quite simple. Basically, I needed to drastically reduce the amount of sugar I eat, eat more protein and increase the amount of omega-3 fatty acids – derived from fish oil – in my diet. These fatty acids are one of the basic building blocks of the brain. The membranes of our brain cells are thought to be about 20 per cent fatty acids and seem crucial for the smooth movement of brain signals. This class of fat is called 'essential' by doctors because our bodies cannot produce it, unlike other nutrients. Only certain elements of our diet can give it to us, specifically seafood, walnuts, leafy greens and flaxseed. 'You clearly have much to gain by increasing your intake of these things, as well as reducing your consumption of carbohydrates and, ideally, eliminating sugars.'

This wasn't going to happen, I told Cox. The idea of giving up a moderate but essential-to-my-mental-wellbeing amount of chocolate and puddings is too bleak. He smiled. 'I understand,' he said. 'Being completely dogmatic about diet is difficult and, in most cases, moderation is the best way.' He gave me a detailed dietary plan as part of his fee, which I said I realistically wouldn't follow (the idea of kippers for breakfast every day makes my throat constrict), but would prefer to derive broad recommendations from. He understood, suggesting that reduced carbohydrate and sugar intake – perhaps including switching to gluten-free products where available – along with increased protein and probiotic foods would make a difference. Oh, and drinking a lot of water. Cox is evangelical about water. In fact, now I remember, every time I saw him he was drinking from a bottle of Evian throughout the appointment. It became a little unnerving at times.

So, I began eating less bread, less rice and less pasta and switching to gluten-free products when I was doing my food shopping. I'd stopped eating red meat a while ago; now I bought more fish (luckily mackerel and sardines are both cheap and among my favourite things to eat) and stocked up on live yoghurt, sauerkraut and miso to up my probiotic intake. Making these changes was not difficult for me as it was largely how I ate anyway, with some tweaks and a bit more discipline in terms of not eating cake or half a loaf of toasted sourdough with jam in one go – although sourdough is thought to be a better option than yeasted bread because of how it's made: with a 'starter' of lactobacillus cultures (a probiotic that benefits your gut) that ferments the dough before baking.

Within a couple of weeks of these changes, which I purposefully implemented in the latter half of my cycle, I *did* notice a difference. I was not nearly as bunged-up-feeling as usual and 'going' a lot more regularly. The cramping, bloating and nausea I had become so accustomed to reduced and, in that crucial week before my period, my dips in mood and bouts of anxiety or panic did not disappear, but felt less pronounced. Perhaps part of this was the placebo effect of actively *doing* something for myself, consciously adapting my life because someone with scientific knowledge told me it might help. It's impossible to tell. More importantly, I don't care. I have stuck with these simple modifications because they seem to help a bit, and that is enough for me.

Forgetting the basics

Although gut ecology is vast and complex, and an expensive test ordered by an expensive clinical nutritionist had identified certain biological factors that justified my symptoms, a lot of what was said was common sense. But a test is never a complete diagnostic tool. If a person has gut-related symptoms, it is a given that their microbiome is skewed.

And the solution, regardless of whether we test positively for dysbiosis, SIBO or yeast overgrowth, is the same: we need to address what caused the imbalance in the first place. We know that the modern diet is loaded with carbohydrates and processed grains. These feed the unhelpful bacteria in the gut. Moderating our intake of these, eating more vegetables than anything else and incorporating probiotic and omega-3-rich foods into our diet has long been recommended as the most healthy approach. We don't need fancy tests to tell us this is probably the best way to eat.

I liked Cox a lot and ended up recommending him to friends with dietary issues, who also found his recommendations beneficial. However, I did not follow his suggestion of further investigations to 'confirm' my findings, such as a month-long test called the 'Rhythm Plus' – also by Geneva Diagnostics – that involved spitting into a sample pot every day so my cortisol levels could be measured and analysed in accordance with my fluctuating hormone levels. This would give a dataset that would identify whether I was more stressed in the latter half of my cycle. I don't need a test to tell me what I already know is true. He also recommended taking some hefty-sounding probiotic pills that may or may not make me feel 'considerably worse' to start with. This, coupled with the steep price, was a no-no for me. I realised how easy it would be to fall into a cycle of constant investigation to identify specific-yet-not-dangerous health 'abnormalities' that would be unlikely to be picked up within the NHS.

It is seductive, the idea of homing in on all these hidden clues about why we might feel a certain way, and I do feel enlightened by what I learned about metabolising oestrogen. However, that the private sector offers all these clever and more targeted-sounding tests and alternatives should also be seen as evidence in itself that it might not be as superior as it seems. Despite its flaws, the NHS offers evidence-based medicine and works to guidelines informed by research as it happens. Sometimes it takes a while to catch up, but it is not, by any means, less sophisticated an entity than the world of private medicine. Perhaps the tests, approaches and remedies offered solely in the private sector are not available elsewhere because there is not enough robust evidence for them to be widely approved. Also, a blood, urine or stool test will only give you information based on your physiology at that time, on that day. It is not a truly representative sample. To get a full picture of a woman's physiology in relation to her menstrual cycle would be nigh on impossible. Even a month's worth of saliva will only give data that reflect *that* month. If she eats differently, encounters a new stressor at work or at home, fights off a virus or infection, travels, takes up a new exercise, starts having sex with a new partner – the variables are endless – the next month's data set could be completely different. As Panay said when I met him, 'tests have a limit to their usefulness'. They give a snapshot. 'We always say, "First treat the patient, not the result."'

Functional medicine forefronts nutrition as the gateway to optimal living, but does not have the same research premise as 'conventional' medicine. Biochemistry is still involved and the outcomes may seem more refined, but the trial process is not the same. Double-blind, placebo-controlled studies are not used; in the eyes of many scientists, this reduces reliability. Like the feeling of safety the wellness industry offers, though, the idea of being heard, seen and treated as an individual with complex individual needs is attractive. But we should be encouraged to remember that, aside from serious gut issues like coeliac disease (an autoimmune condition) or true lactose intolerance, only picked up in blood tests, so much of our approach to our health can, and should, make use of common sense.

'In both the private and public sector, doctors are often guilty of forgetting the basics,' says Dr Elaine McQuade, a GP on the team at the Marion Gluck Clinic, a private Harley Street clinic that specialises in the use of bio-identical hormone replacement therapy (BHRT) to, they say, 'restore and maintain optimal health'. 'We need to get back to a place where people are reminded that sleeping well, exercising and having a good diet – things we can all control – are the absolute foundations for good health,' she says.

Bio-identical hormones

I was surprised to hear the sentiment of common sense being so passionately pushed by a doctor at a Harley Street clinic. I had come to the Marion Gluck Clinic after hearing the term 'bio-identical' a lot in relation to hormone treatments while researching this book. You often hear famous women talking about it and invariably you end up at the same place: the Marion Gluck Clinic. On the website there is a section titled, 'Who do A-Listers see to beat the menopause?' Author Jeanette Winterson wrote a long piece for the *Guardian* about how bio-identical hormones prescribed by Gluck changed her life after a premenopausal breakdown.

> Over the last few years I have been trying to find a medical model responsive to the fact that our bodies are not machines made of faulty parts but a process in constant movement and change [she writes]. And I have been looking for medical practitioners able to discuss the mind–body balance as a synergy, not a confusing, frightening war.[84]

Wary of the existing options of antidepressants or HRT, Winterson saw a series of practitioners, including a nutritionist, before being directed to Gluck's clinic. After a series of blood tests and a bespoke BHRT regime, 'I feel at home in my body again,' she wrote. She has since become an ambassador of sorts for the clinic, interviewing Gluck on camera for the site. When I interviewed Gillian Anderson and her friend the journalist Jennifer Nadal, around the release of their book *We: A Manifesto for Women Everywhere*, Anderson talked about how, when she became perimenopausal, she felt like someone else had 'taken over' her brain. She said bio-identical hormones had been a great help. So what are they and why are they better?

According to the literature, bio-identical hormones have the same chemical structure as the naturally occurring hormones which are produced in the body, unlike the synthetic versions such as Premarin, which is the oestrogen-replacement HRT drug that can help with menopausal symptoms like depression, anxiety, hot flushes, tiredness, vaginal dryness and reduced libido. The Marion Gluck Clinic say that bio-identical hormones are derived from a substance called diosgenin, which is 'sourced from Mexican yams'. They are, it says, '100 per cent identical in chemical structure to our own hormones, and their effects and benefits replicate them closely. Bio-identical hormones can therefore be very beneficial to patients who have a hormone imbalance or have previously tried other hormonal medications, including HRT, and experienced unwanted effects.' In theory, as the clinic states, as well as menopause BHRT could help to treat other hormone-related conditions like PMS, postnatal depression, endometriosis, PCOS and, in men, andropause (a gradual decrease in testosterone which can become symptomatic in the forties and fifties).

HRT horses

What's wrong with Premarin? The answer depends on your general interest in where medicine comes from, and your stomach for animal welfare. In her *Guardian* piece, Winterson suggests to 'pour yourself a stiff drink' before googling it. Premarin is the brand name for conjugated equine estrogens (CEEs). It is the HRT drug used to help treat symptoms of menopause and prevent osteoporosis, as well as helping to treat prostate cancer in men and breast cancer in men and women. 'Premarin' is short for pregnant mare's urine. According to information from organisations like PETA, some mares can be kept in permanent indoor stalls, with a heavy rubber bag attached to their groin 24/7. They can be kept thirsty to make their urine richer. When they give birth the mare's foals are often taken away and they are re-inseminated as soon as possible to begin another eleven-month cycle. When the mares' bodies eventually break down, many are assigned to the horse-meat market. Winterson says if you are an ethical vegetarian on HRT you 'might as well eat foie gras every day'.

This may be true. But plenty of medications humans take every day are derived from animals. Many people with thyroid problems take levothyroxine, the synthetic thyroid hormone replacement drug. However, some choose to take a 'natural' thyroid replacement (only available privately) known as desiccated thyroid hormone (DTH). Currently, desiccated thyroid extract is made from pig thyroids. Heparin, an incredibly important blood-thinning drug used to treat and prevent deep vein thrombosis (DVT) and pulmonary embolism, and also used in the treatment of heart attacks and unstable angina, is derived from the mucosal tissues of pig intestines and cow lungs. If we were in a hospital resuscitation

226

room having a heart attack, would we reject a blood-thinner because it has come from a pig gut? Maybe that doesn't work as an analogy, given that heparin is derived from already slaughtered animals, but there is a point to be made about our selective moral mathematics about where things come from.

The lure of precision

I was dubious about bio-identical hormone replacement therapy because it just seemed too good to be true. However, I was curious about what somewhere like the Marion Gluck Clinic might offer me, which is why I found myself in a room with Dr McQuade. Having made contact with the clinic through their media person, I was informed that any initial consultation and tests would be 'taken care of'. I made clear that my reporting would be reflective of my experience, which might not be wholly positive, and they were sanguine about this. I liked McQuade immediately. She had a great handshake, confident, broad movements and a soft Northern Irish accent. McQuade explained that women go through a lifecycle of hormonal fluctuations from puberty to menopause. Fluctuations, as we know, are completely normal and for the most part our body is in a constant state of readjustment to change. 'The problem is that the body cannot always rebalance itself,' she says. 'In the modern world, things like poor nutrition, toxins in the environment and increased stress can have an effect on the body's ability to rebalance itself.' As the theory goes, a hormonal imbalance can affect energy, mood, fertility, sleep, sex and more. Gluck says in her interview with Winterson that bio-identical hormones 'can be used at every stage of life – not only during the menopause. A mental breakdown does not happen only in the head.'

Quite right. So, suppose I wanted to see whether BHRT would help with my PMS. What would happen? McQuade says a comprehensive hormone profile would be established via a blood test in the latter half of my cycle, to help see whether there is an imbalance. If there was, the relevant bio-identical hormones could be prescribed at a dose tailored directly to my needs. My curiosity was piqued. I said I'd be interested in what a blood test showed. A week later, on day 21 of my cycle and with a form from the clinic, I went to an independent pathology lab round the corner from Harley Street that was more hotel lobby than clinic: all steely surfaces and chunks of bright-coloured soft furniture. It was unnerving. A few days later, the clinic emailed me to say I could come in and discuss my results with McQuade. The hormone profile test had looked at my thyroid hormone levels, follicle-stimulating hormones (FSH), oestrogen, progesterone, testosterone, Vitamin D and DHEA. I'd never heard of this last substance, a hormone called dehydroepiandrosterone that is produced in the adrenal gland and helps produce other hormones – such as oestrogen. Natural DHEA levels peak in early adulthood, then slowly fall as you age, and are associated with mood, energy and bone health. So what was wrong with my hormones? 'Nothing,' said McQuade.

I don't know why, but I was surprised. So was she.

'Based on what you told me, I was expecting to see oestrogen dominance,' she said, 'but everything is within the normal ranges associated with this part of your cycle.' My thyroid hormone levels were at the lower end of normal, she said, but it didn't warrant treatment. Interestingly, about ten years ago, a blood test showed I did have an underactive thyroid. I was put on levothyroxine and took it for around two years, having blood tests every six months to check I was within the normal range. When I went the US on a work trip I forgot to pack the tablets and, although slightly concerned about what might happen, just stopped taking them when I got back. Every blood test I have had thereafter has shown that I am in the normal range. I told McQuade this and she said that, sometimes, the thyroid will show up as being temporarily 'underactive' after a period of illness. Thinking back, I'd had a horrific bout of tonsillitis. Maybe that did it. I wondered if the lack of 'oestrogen dominance', as she put it, could have been a result of the dietary changes I'd made after seeing Cox. She asked about my bowel movements and I said I'd been less constipated. 'It could be, then. If you're hanging on to waste in the gut for too long, there is going to be more oestrogen in your system than there should be. But it would be impossible to say for certain.' I asked what treatment, if any, she might recommend to help with my PMS. 'Not BHRT.'

Again, I was surprised. A Harley Street doctor not finding a way to offer me some kind of treatment seemed peculiar.

'You are thirty-three years old. We don't want to be giving you hormone preparations if we don't have to. I think, for you, looking at the stressors in the rest of your life and focusing on the basics is the way to go.'

My brief experience with the Marion Gluck Clinic was extremely positive. I felt welcomed, heard and treated as an individual by a woman who seemed genuinely interested in my wellbeing and had all the time in the world for my questions. This may have been because the clinic's public relations staff had facilitated the appointment. McQuade may well have been briefed, or advised, to give me the best experience possible. However, it does not surprise me at all that women travel to the clinic from across the world. Of course, this is only women who have money. Were the fees not waived for me by the media team in light of coverage, there is no way on earth I would be able to afford the cost. The initial consultation costs £290. Follow-up appointments are £130, whether in person or on the phone. A nutritional consultation is £100. Blood tests are charged separately. The female hormone profile test I had would have cost me £295, plus £90 for the thyroid profile. Repeat prescriptions cost £25. These are prohibitive amounts for most normal women. And all for a treatment that, although many seem to deem it life-changing, has been rightfully questioned in 'conventional' science.

'Bio-identical' is not a scientific phrase, it is a marketing one. The 'bio' part implies superiority over the federally approved hormone treatments. The two naturally occurring sex hormones made by the ovaries and used to help treat menopausal symptoms are estradiol and progesterone. Both of these hormones are available in FDA- and NICE-approved, pharmaceutical-grade preparations. They come with a significant body of evidence that outlines the benefits and risks. Not so with BHRT. This is because the products have not been through the same rigorous scientific testing. This doesn't mean they don't work for a lot of women, but there is no good evidence to suggest that these less well tested hormones will be any 'better' for us than what is already out there. A BHRT cream applied to the wrists twice daily is not a treatment for a woman's self-esteem. Gluck counters this point in her interview with Winterson by saying that, 'as a doctor with thousands of hours of clinical experience, working with women across the spectrum of age and health and fertility, I know that BHRT can improve a person's total wellbeing. And wellbeing has mental, as well as physical, benefits.'

We know that society has a lot of funny ideas about older

women. Once women head towards the end of their repro-ductive years, the gaze of oppression shifts from our fecundity to our lack thereof; our barrenness, our lack of vigour. Older women often struggle to value themselves. BHRT may seem like a safer, more sensible and softer option, but wherever we are on the reproductive continuum, we need to think bigger than our biology. We also need to think about what women are finding in places like the Marion Gluck Clinic that they feel they haven't found before: time, attention, compassion and multi-dimensional approaches to multi-dimensional problems. A holistic approach to our health and wellbeing is a very seductive thing.

Society has always tried to contain women's pain, pleas-ure and so-called excesses. It was worse back in time, sure, but however the systems that surround our bodies may have changed, what feels bad or uncomfortable to us *now* is valid and relevant. We do not have to qualify our pain or discomfort based on what might have happened to us in the past. Our truths, biological and emotional, have been derided, oppressed and repackaged back to us as different kinds of pathology that we can, should, fix. Tyrannies of female ideals, a direct product of the way the patriarchy has tried to contain us, have created this notion that a woman 'should' be stable, calm, kind, soft and nurturing and that movement from this is a sign of trouble. It has entered our entire species. It is in us all, somewhere far down in our consciousness, and makes us constantly analyse our emotional responses to the world in case someone doesn't like what they see or hear. We fear our anger. We fear what our bodies do and what comes out of them because it all points to that terrible thing: *excess*. We are wriggling for vocal, bodily, emotional freedom in a world that doesn't really want to allow it to us yet. But we have to keep wriggling.

Over the course of writing this book, more and more I began to see that the only way to really live with my moments of cyclical despair was the hard way: to try to reconceptualise it. I continue to work with a fantastic psychologist who helped me make sense of how past trauma, pain and my sense of myself had informed the way I see who I am in the present, and the way I relate to other people. Given that so much of my PMS anguish comes from the way I feel I'll be seen, this has been very important.

Just like any hormone-related process a woman goes through in her life, from menarche to menopause, PMS is a very complex experience – and I hope I have demonstrated just how complex the relationship between body and mind is. As human beings, rhythm is part of our deep core. Breathing, pumping, blinking, digesting, thrusting, contracting; these are all actions of the body that keep us alive, and keep our species going. As women, we have in menstruation another rhythm unique to us. But while our biology and what is happening inside us is important, we cannot act as if the body is an anatomical drawing floating in space; that it is nowhere. Context is as integral a part of how we experience our hormonal fluctuations as the physical matter that makes us.

When I spoke to Hustvedt, she said she believed that part of the problem with understanding why some women find these processes difficult is that we have spent so long placing women in nature and men outside nature. 'We know this is not true,' she said. 'Men are born and they die. Women are born and they die. We are all natural creatures. If the models of how we research and understand hormones in women changed, if there wasn't such an endless search for biological "reasons" and much more of a social approach about how women are positioned in the world, I think it would change a lot.'

Finally

I agree with Hustvedt wholeheartedly, yet the day-to-day reality of our lives in a world still so tyrannical about how women should be, or be seen, means it can be hard to feel like we have any power to effect change. Perhaps, though, to find a more comfortable way through the macro picture we can try and do something with the micro one. We do so much to rid ourselves of the idea that we are slaves to the moon cycle. What would it look like if we could see the way our moods and behaviour correspond with different parts of our cycle – or indeed any process involving our reproductive components – as a *part* of us, rather than a machine going into 'monster' mode? What if we could work towards dismantling the self-blame and stop othering these parts of who we are?

The mythology that has surrounded women's bodies since time immemorial, positioning our natural processes like menstruation as dirty and dangerous, does not make this an easy ride. As I said much earlier on, it is very easy to dismiss mythology now we live in an age of science, technology and supposedly clear-cut 'explanations' for why we feel a certain way when something is happening with our reproductive body. But the truth is that, because the science is still inexact when it comes to our hormones, there is no reliably clear-cut explanation available. If we feel sad, anxious and irritable before our period, it is not 'just' our hormones making us feel that way. Life is so rarely black-and-white; it is shades of grey.

Our physical and mental selves are interrelated in ways we can barely get our minds around. That is why there is such variation between women when it comes to our experiences of things like menstruation, pregnancy and menopause. There may be times when we have more extreme hormonal fluctuations and really feel like there's a 'chemical' nature to our mood (I know I've felt this), but those may also be times when things are difficult in our lives anyway. A perfect storm. How our bodies react to different levels of hormones may also be to do with how well we are at that particular time generally, both immunologically and psychologically. How we react to someone we feel is being an arse when we are premenstrual might just be because they really are being an arse. It might not be because we're hormonally unhinged.

If we are open to the possibility that the modern-day surveillance of the reproductive body has transformed all those myths into medical, legal and scientific 'truths', we can really start to think about why the wider experience of being a woman – particularly a woman in some kind of distress – comes with so much exhausting ideology that feeds into even our own private, individual experiences. It feeds our inner voice. I have spent so much time throughout my adult life questioning whether the mental potholes I can fall into mean my nature is faulty. When those potholes appear in line with particular points in my cycle I question even further, believing there really *is* an inevitability to my distress as a woman. My quest for peace is non-linear and ongoing. However, when I reached a crossroads in trying to treat or quell my premenstrual anguish, when the existing treatment options effectively ran out or became much more drastic, I actually did begin to think differently. I never thought that could ever happen, but it did. I looked right at the idea of there being nothing more a doctor could do and wondered what *I* could do. What had I been missing? Where might I get if I stopped asking *why* and stopped so readily analysing my mood, thoughts and behaviour?

Part of this process involved a kind of mental panning out

and thinking about the world I inhabit. What came before me? What orbits my corporeal self that could be affecting how I have the capacity to feel while my biochemistry is changing? I am still realising that so much of the torment lies in fearing how I will be perceived. Over the course of researching and writing this book, I have better understood not just the mechanics of my body and what might be happening within it when I feel bad, sad or mad, but the mechanics of a society that has, for so long, in so many ways, tried to contain female excess. I hope reading this book may have done the same for you.

The meaning of our emotion and our pain, in whatever form it takes, has historically been decided for us. In the enclaves of our subconscious minds we wrestle with how much of our inner world we should show, in case we aren't properly heard or in case what comes out makes us somehow less than – or worse than – what history has told us we should be. But we all have the capacity to relearn ourselves and feel more in sync with the bodies we inhabit. We should feel able to ask questions of the systems that surround us, educate ourselves and ask for more – or better – if we're dissatisfied.

Virgil was right: woman *is* ever changeable. But it is in our capacity to inquire, analyse, look around and ask for better that we are most powerful.

Acknowledgements

Thank you to the women in my life who make me feel seen and sick with laughter, and have particularly over the course of writing this book: Kate Merry, Nell Frizzell, Hayley Campbell, Eva Wiseman, Pip Hartle, Alexandra Heminsley, Alice Grier and Charlotte Mendelson. Also to Peggy: you have four legs and a tail but you are my great love. Special thanks to the Society of Authors for their generosity in helping me finish this book.

Notes

Part One

The female animal

1. 'Thanksgiving In Mongolia', Ariel Levy, *The New Yorker*, 18 November 2013
2. *The Stressed Sex: Uncovering the Truth About Men, Women, and Mental Health*, Daniel Freeman, Jason Freeman, Oxford University Press (2013)

Part Two

Anatomy

3. 'The hidden symbols of the female anatomy in Michelangelo Buonarroti's ceiling in the Sistine Chapel', Deivis de Campos et al., *Clinical Anatomy*, October 2016, Volume 29, Issue 7, 911–16, https://onlinelibrary.wiley.com/doi/abs/10.1002/ca.22764
4. 'This "Sistine code" theory is daft. Michelangelo is not a feminist hero', Jonathan Jones, *Guardian*, 2 September 2016
5. *Leonardo da Vinci: Anatomist*, Martin Clayton and Ron Philo, Royal Collection Trust; 1st edition (2012)
6. 'Leonardo da Vinci's Embryological Drawings of the Fetus', Hilary Gilson, *Embryo Project Encyclopedia*, 19 August 2008, ISSN: 1940-5030, http://embryo.asu.edu/handle/10776/1929

Puberty

7. Information retrieved from: https://www.nhs.uk/live-well/sexual-health/stages-of-puberty-what-happens-to-boys-and-girls

8. 'The timing of normal puberty and the age limits of sexual precocity: Variations around the world, secular trends, and changes after migration', A. Parent, G. Teilmann, A. Juul, N. E. Skakkebaek, J. Toppari, J. Bourguignon, *Endocrine Reviews,* 1 October 2003, Volume 24, Issue 5, 668–693, https://doi.org/10.1210/er.2002–0019

9. 'The Physiology of Menstruation. In: Dysmenorrhea and Menorrhagia', R. P. Smith (2018), Springer, Cham, https://link.springer.com/chapter/10.1007/978-3-319-71964-1_1

Eggs and fertility

10. 'Human Ovarian Reserve from Conception to the Menopause', W. H. B. Wallace, T. W. Kelsey (2010), PLoS ONE 5 (1): e8772, https://doi.org/10.1371/journal.pone.0008772

11. 'Women lose 90 per cent of "eggs" by 30', Richard Alleyne, *Daily Telegraph,* 27 January 2010

12. 'Women are turning to birth control smartphone apps for a reason', Dawn Foster, *Guardian,* 24 July 2018

13. 'Published analysis of contraceptive effectiveness of Daysy and DaysyView app is fatally flawed', Chelsea B. Polis, *Reproductive Health* 15 (2018), 113, https://reproductive-health-journal.biomedcentral.com/articles/10.1186/s12978-018-0560-1

14. 'The role of sexual conflict in the evolution of facultative parthenogenesis: a study on the spiny leaf stick insect', Nathan W. Burke, Angela J. Crean, Russell Bonduriansky (2015), *Animal Behaviour* (101), 117–27, https://doi.org/10.1016/j.anbehav.2014.12.017

15. 'Association Between Biomarkers of Ovarian Reserve and Infertility Among Older Women of Reproductive Age', A. Z. Steiner, D. Pritchard, F. Z. Stanczyk et al. (2017), *Journal of the American Medical Association*, 318 (14): 1367–76, doi:10.1001/jama.2017.14588

16. 'Egg Freezing in Fertility Treatment: Trends and Figures 2010-2016', The Human Fertilisation and Embryology Association, 20 December 2018, https://www.hfea.gov.uk/media/2656/egg-freezing-in-fertility-treatment-trends-and-figures-2010-2016-final.pdf

Blood

17. *Carrie*, Steven King, Doubleday Press (1974)

The happiness fantasy

18. *An Essay Concerning Human Understanding*, John Locke, Penguin Classics; revised edition (1 February 1998)
19. 'Cranes in the Sky' by Solange Knowles (songwriters Solange Knowles, Raphael Saadiq), 5 October 2016, by Saint Records and Columbia Records

The cycle: a vital sign

20. 'Variation of the Human Menstrual Cycle Through Reproductive Life', A. E. Treloar, R. E. Boynton, B. G. Behn, B. W. Brown, *International Journal of Fertility*, 1 January 1967, 12 (1 Pt 2), 77–126.
21. 'Chlorination by-products in drinking water and menstrual cycle function', G. C. Windham, K. Waller, M. Anderson, L. Fenster, P. Mendola, and S. Swan (2003), *Environmental Health Perspectives*, 111 (7), 935–41, https://www.ncbi.nlm.nih.gov/pmc/articles/PMC1241528/
22. 'Long-term exposure to trihalomethanes in drinking water and breast cancer in the Spanish multicase-control study on cancer (MCC-Spain)', F. B. Laia et al., *Environment International*, March 2018, Volume 112, 227–34, https://doi.org/10.1016/j.envint.2017.12.031

Menstrual phase (days 1 to 5)

23. 'Prostaglandins and Inflammation. Arteriosclerosis, Thrombosis, and Vascular Biology', E. Ricciotti and G. A. FitzGerald (2011), 31 (5), 986–1000, http://doi.org/10.1161/ATVBAHA.110.207449
24. 'Burden of migraine related to menses: results from the AMPP study', *The Journal of Headache and Pain*, Pavlović, J. M., Stewart, W. F., Bruce, C. A. et al. (2015) 16: 24, https://doi.org/10.1186/s10194-015-0503-y

Luteal or 'secretory phase' (days 15 to 28)

25. Information retrieved from: http://www.pms.org.uk/about
26. 'Change in women's eating habits during the menstrual cycle', Ines Kammoun, Wafa Ben Saâda, Amira Sifaou, Emna Haouat, Hajer Kandara, Leila Ben Salem and Claude Ben Slama (2016), *Annals of Endocrinology* (*Annales D'Endocrinologie*, English Edition), Volume 78 (1), 33–7
27. 'The menstrual cycle and the skin', R. S. Raghunath, Z. C. Venables, G. W. M. Millington (11 February 2015), *Clinical and Experimental Dermatology*, March 2015, Volume 40, Issue 2, pp. 111–15 doi: https://doi.org/10.1111/ced.12588

The thyroid

28. 'Graves Disease is Associated with Endometriosis: A 3-Year Population-based Cross-sectional Study', J. S. Yuk, E. J. Park, Y. S. Seo, H. J. Kim, S. Y. Kwon and W. I. Park (2016), *Medicine*, 95 (10), e2975

Part Three

29. *Managing the Monstrous Feminine: Regulating the Reproductive Body*, Jane M. Ussher, Routledge (2005)

Leaky

30. '#MenstruationMatters: Taboo around menstruation causing women shame, researcher says', Kellie Scott, 25 May 2016, http://www.abc.net.au/news/2016-05-25/taboo-around-menstruation-causin g-women-shame/7445378
31. 'Silence and the History of Menstruation [online]', Carla Pascoe (2007), *The Oral History Association of Australia Journal*, No. 29, 28–33
32. *Powers of Horror: An Essay on Abjection (European Perspectives)*, Julia Kristeva, Columbia University Press (1984)

Hysteria: 'an animal within an animal'

33. *Women and Society in Greek and Roman Egypt: A Sourcebook*, Jane Rowlandson, Cambridge University Press (2009)

Hunting witches

34. 'Harvey Weinstein Is My Monster Too', Salma Hayek, *New York Times*, 12 December 2017
35. 'Yes, This Is a Witch Hunt. I'm a Witch and I'm Hunting You', Lindy West, *New York Times*, 17 October 2017

Reputation

36. 'Actors are lining up to condemn Woody Allen. Why now?', Hadley Freeman, *Guardian*, 3 February 2018

Don't be so sensitive

37. 'Princess Diana Was As Mad As Any Other Woman', Sophie Heawood, VICE.com, 14 March 2014

Exploding

38. '"It's playful and anarchic": The cast of Killing Eve on Phoebe Waller-Bride's killer thriller', Rebecca Nicholson, *Guardian*, 15 September 2018

39. 'Fleabag star Phoebe Waller-Bridge on female anger, emotional honesty – and fancying Barack Obama', Elizabeth Day, *Daily Telegraph*, 7 July 2016

Surveillance

40. Information retrieved from: https://unesdoc.unesco.org/ark:/48223/pf0000226792

41. 'Richard Maurice Bucke, M.D. 1837–1902: The Evolution of a Mystic', Cyril Greenland (1966), *Canadian Psychiatric Association Journal*, 11 (2), 146–54, https://doi.org/10.1177/070674376601100212

Bad clitoris!

42. *On Surgical Diseases of Women*, Isaac Baker Brown, John W. Davies (1861)

43. *The Technology of Orgasm: "Hysteria," the Vibrator, and Women's Sexual Satisfaction*, Rachel P. Maines, Johns Hopkins University Press (2001)

44. *Nerve-Vibration and Excitation as Agents in the Treatment of Functional Disorder and Organic Disease*, J. Mortimer Granville, Forgotten Books (2018)

45. *Three Essays on the Theory of Sexuality: The 1905 Edition*, Sigmund Freud, Verso (2017)

'Returning' to the sexual body

46. 'The Space in Between Naomi Wolf's "Vagina: A New Biography."', Ariel Levy, *New Yorker*, 10 September 2012

47. *Vagina: A New Biography*, Naomi Wolf, Virago (2013)

48. *The Sexually Adequate Female*, Frank S. Caprio, The Citadel Press (1964)

49. 'Anatomy of the Clitoris', H. E. O'Connell, K. V. Sanjeevan, J. M. Hutson (2005), *The Journal of Urology*, 174 (4 Pt 1): 1189–95

Nipple sadness

50. 'Women's Clitoris, Vagina, and Cervix Mapped on the Sensory Cortex: fMRI Evidence', Barry R. Komisaruk, Nan Wise, Eleni Frangos, Wen-Ching Liu, Kachina Allen, Stuart Brody (2011), *The Journal of Sexual Medicine*, Volume 8, Issue 10, 2822–30, https://doi.org/10.1111/j.1743-6109.2011.02388.x

Part Four

Personality

51. 'How universal is the Big Five? Testing the five-factor model of personality variation among forager–farmers in the Bolivian Amazon', M. Gurven, C. Von Rueden, M. Massenkoff, *Journal of Personality and Social Psychology*, Volume 104 (2), February 2013, 354–70
52. 'The Mysterious Popularity Of The Meaningless Myers-Briggs (MBTI)', Todd Essig, www.forbes.com, 29 September 2014

Softening plaster

53. *The Principles of Psychology: Volume One*, William James, Dover Publications Inc.; New edition (2000)

Imbalance

54. 'The Culture of Prozac', Newsweek Staff, *Newsweek*, 2 June 1994
55. 'Bad Mothers and Single Women: A Look Back at Antidepressant Advertisements', Katherine Sharpe, *Huffington Post*, 11 August 2012

PMS and neurotransmitters

56. Information retrieved from: https://cks.nice.org.uk/premenstrual-syndrome
57. 'Sex hormones affect neurotransmitters and shape the adult female brain during hormonal transition periods', C. Barth, A. Villringer, and J. Sacher, *Frontiers in Neuroscience*, 20 February 2015, 9, 37, doi:10.3389/fnins.2015.00037

Gender differences in mental health: not that simple

58. Information retrieved from: http://www.who.int/mental_health/prevention/genderwomen/en/
59. 'Hormone-specific psychiatric disorders: do they exist?', M. Altemus (2010), *Archives of Women's Mental Health*, 13 (1), 25–6

Formal recognition

60. 'Should Severe Premenstrual Symptoms Be A Mental Disorder?', Amy Standen, 21 October 2013, www.NPR.org

61. 'British Legal Debate: Premenstrual Tension and Criminal Behaviour', *New York Times*, 19 December 1981, obtained through *New York Times'* print archive, https://www.nytimes.com/1981/12/29/science/british-legal-debate-premenstrual-tension-and-criminal-behavior.html

Diagnosis: not for everyone

62. 'Experiencing mental health diagnosis: a systematic review of service user, clinician, and carer perspectives across clinical settings', A. Perkins, J. Ridler, D. Browes, G. Peryer, C. Notley, C. Hackmann (2018), *The Lancet*, 5 (9), 747–64
63. 'Mental health labels can save lives. But they can also destroy them', Jay Watts, *Guardian*, 24 April 2018

Borderline personality disorder: a bad name for real suffering

64. 'Responses of mental health clinicians to patients with borderline personality disorder', R. A. Sansone, L. A. Sansone (2013), *Innovations in Clinical Neuroscience*, 10 May 2013, (5–6), 39–43
65. *The Body Keeps the Score: Brain, Mind, and Body in the Healing of Trauma*, Bessel Van Der Kolk, Penguin (2015)

The weight of trauma

66. Information retrieved from: https://www.cdc.gov/violenceprevention/childabuseandneglect/acestudy/index.html

Conversion

67. *It's All In Your Head: Stories from the Frontline of Psychosomatic*, Suzanne O'Sullivan, Vintage (2016)
68. *The Shaking Woman or A History of My Nerves*, Siri Hustvedt, Sceptre (2011)
69. 'The Shaking Woman or a History of My Nerves by Siri Hustvedt', Hilary Mantel, *Guardian*, 30 January 2010

Part Five

Pain

70. *Illness as Metaphor and AIDS and Its Metaphors*, Susan Sontag, Penguin Classics (2009)

71. 'The Girl Who Cried Pain: A Bias Against Women in the Treatment of Pain', Diane E. Hoffmann, and Anita J. Tarzian (2001), *Journal of Law, Medicine & Ethics*, Volume 29, 13–27, http://dx.doi.org/10.2139/ssrn.383803

72. Information retrieved from: https://www.bsge.org.uk/guidelines/

73. 'Gender Disparity in Analgesic Treatment of Emergency Department Patients with Acute Abdominal Pain', Esther H. Chen, Frances S. Shofer, Anthony J. Dean, Judd E. Hollander, William G. Baxt, Jennifer L. Robey, Keara L. Sease, Angela M. Mills (2008), *Official Journal for the Society of Academic Emergency Medicine*, 15 (5), 414–18

74. 'Period pain can be "almost as bad as a heart attack." Why aren't we researching how to treat it?', Olivia Goldhill, *Quartz*, 15 February 2016

75. 'Microglia in Physiology and Disease', Susanne A. Wolf, H. W. G. M. Boddeke, Helmut Kettenmann (2017), *Annual Review of Physiology* 79 (10), 619–43 https://doi.org/10.1146/annurev-physiol-022516-034406

Due scepticism

76. Information retrieved from: https://www.endometriosis-uk.org

77. Information retrieved from: https://newsroom.heart.org/news/men-more-likely-to-receive-bystander-cpr-in-public-than-women

Birth stories and pain thresholds

78. 'Childbirth stories are the stuff of life. We should share them', Eva Wiseman, *Observer Magazine*, 8 April 2018

79. 'Instead of judging women who want a C-section, why not listen?', Rebecca Schiller, *Guardian*, 21 August 2018

80. Information retrieved from: http://www.who.int/news-room/fact-sheets/detail/maternal-mortality

A word on wellness

81. Information retrieved from: https://globalwellnessinstitute.org/press-room/statistics-and-facts

82. 'Women are flocking to wellness because modern medicine still doesn't take them seriously', Annaliese Griffin, *Quartz*, 15 June 2017

How to treat a problem that is neither 'just hormones' nor 'all in the mind'

83. 'Were there evolutionary advantages to premenstrual syndrome?', M. R. Gillings (2014), *Evolutionary Applications*, 7 (8), 897–904

Bio-identical hormones

84. 'Jeanette Winterson: can you stop the menopause?', Jeanette Winterson, *Guardian*, 11 April 2014